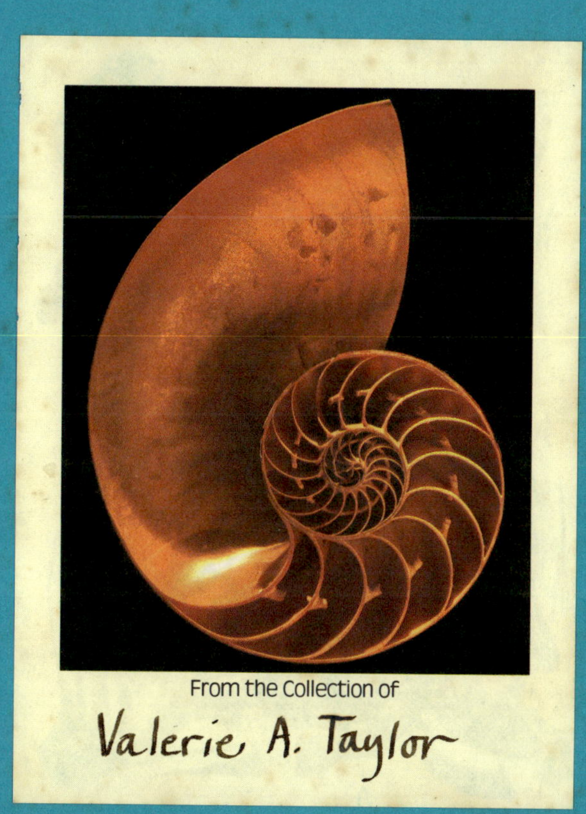

The Noosa Story

Australians really love the sea and the coast. The trouble is
that there are enough of them — or there will be soon —
to turn all of our reachable coastline into a sprawling
suburbia . . . . The preservation of landscape values is an
essential part of wise coastal development.

Francis Ratcliffe, O.B.E., 'The Coast —
Post-war Changes and Overall
Conservation Problems', *Conservation of
the Australian Coast*, Special Publication
No. 7 (Australian Conservation
Foundation, 1972)

The Noosa Story

A study in unplanned development

Nancy Cato

The Jacaranda Press

First published 1979 by
THE JACARANDA PRESS
65 Park Road, Milton, Qld
9 Massey Street, Gladesville, N.S.W.
90 Ormond Road, Elwood, Vic.
303 Wright Street, Adelaide, S.A.
4 Kirk Street, Grey Lynn, Auckland 2, N.Z.

Typeset in 10/11 pt Paladium

Reprinted 1980

Second edition 1982

Printed in Hong Kong

© Nancy Cato 1979, 1982

National Library of Australia
Cataloguing-in-Publication data

Cato, Nancy, 1917–
    The Noosa story.

    2nd ed.
    Previous ed.: Milton, Qld.: Jacaranda, 1979.
    ISBN 0 7016 1681 4.

    1. Regional planning — Queensland — Noosa Heads
    Region. 2. Noosa Heads Region (Qld.) — History.
    I. Title.

711'.3'099432

Cover photo: *The Boiling Pot, foreshore of the Noosa
National Park, with pandanus precariously rooted in
the rocks. (Photo: Lin Martin.)*

CONTENTS

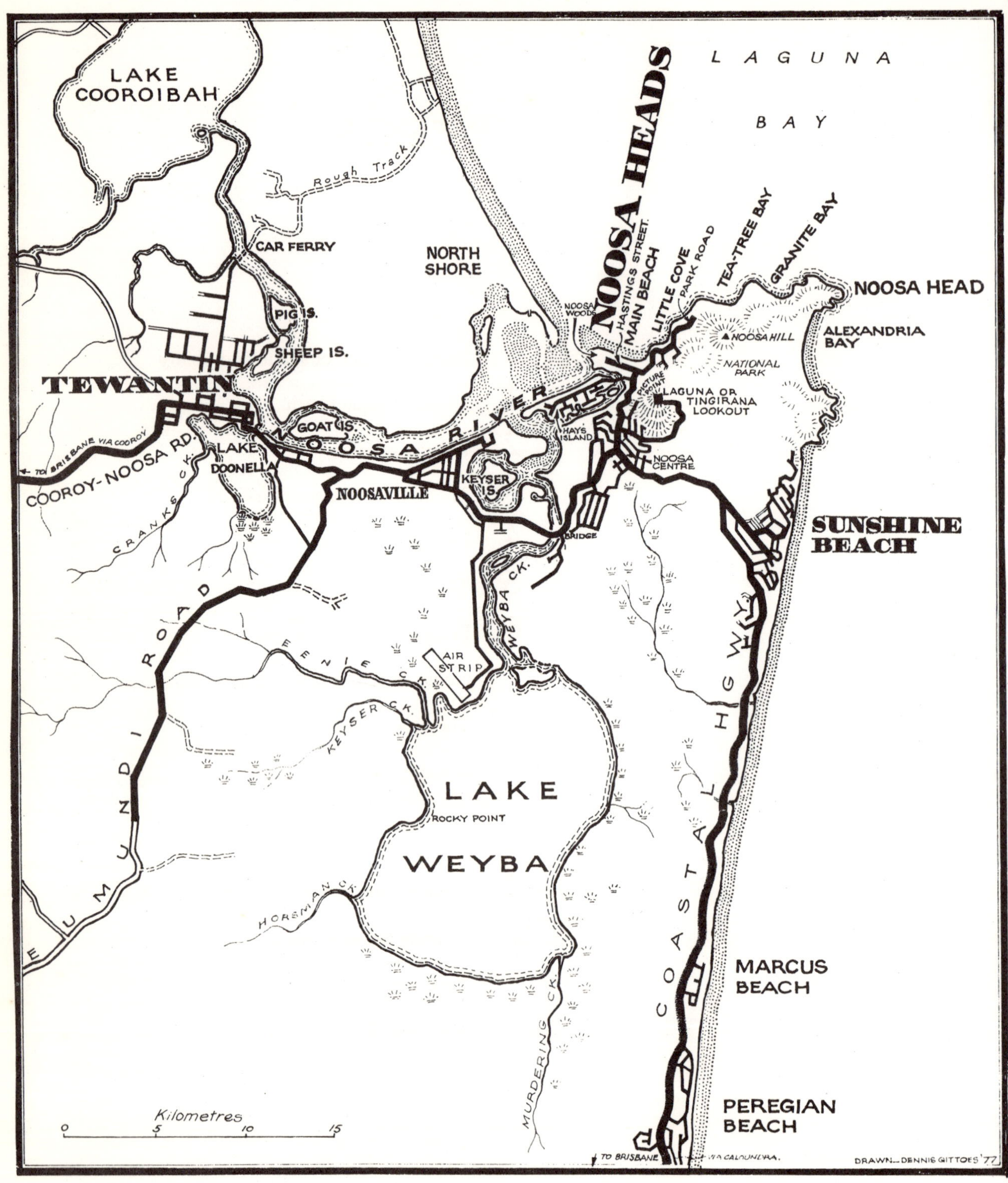

The Noosa estuary. (Drawn by Dennis Gittoes.)

PROLOGUE Noosa: a place of shade and plenty

When I think of what Noosa was like when I was a boy, and what it is now, I could howl with rage and frustration.

Dr C. B. Christesen, O.B.E., F.A.S.A., former editor of *Meanjin Quarterly*

From the top of Motel Hill, as you come over the last rise above Noosa Heads beach, the great curve of Laguna Bay lies sparkling like blue metalassé within the arm of the north shore with its coloured sands. Here, legend says, the rainbow serpent died, and gave his colours and his shape to the earth. On a clear day you can see the end of his broken tail at Double Island Point, sixty kilometres away to the north-west.

The bay, framed between giant cypress and hoop pines with their resinous scent, has an almost Mediterranean air. The wild waves of the Pacific come round the headland in a pattern of ordered curves, the famed 'point break' of the Noosa surf. A necklace of small coves — Little Cove, Tea Tree Bay, Granite Bay — strung out around the headland, provides safe swimming, and the walking track that connects them leads on to Alexandria or Wineglass Bay. All are contained in the 333 hectares of the Noosa National Park.

The rocky headlands ensure a mild and pleasant climate. The northern and eastern sun warms the beach, while the cold south-easterlies pass harmlessly overhead. A wild surf around the corner at Sunshine Beach becomes a steep, even swell at Noosa Heads. When the Pacific is calm during the off-shore westerlies of winter, Laguna Bay looks like a beautiful blue lake.

Noosa Heads is almost entirely surrounded by water. A string of tidal lakes stretches northward, connected by the Noosa River. Blue volcanic hills ring the horizon in isolated peaks.

Little wonder that it has become a tourist mecca and retirement haven for refugees from the cold south. In the southern states of Australia, far from Queensland, you have only to mention that you live at Noosa and someone will say, 'Oh, Noosa! I believe it's a beautiful place', or 'Noosa Heads! That's the place for surf and sandflies.'

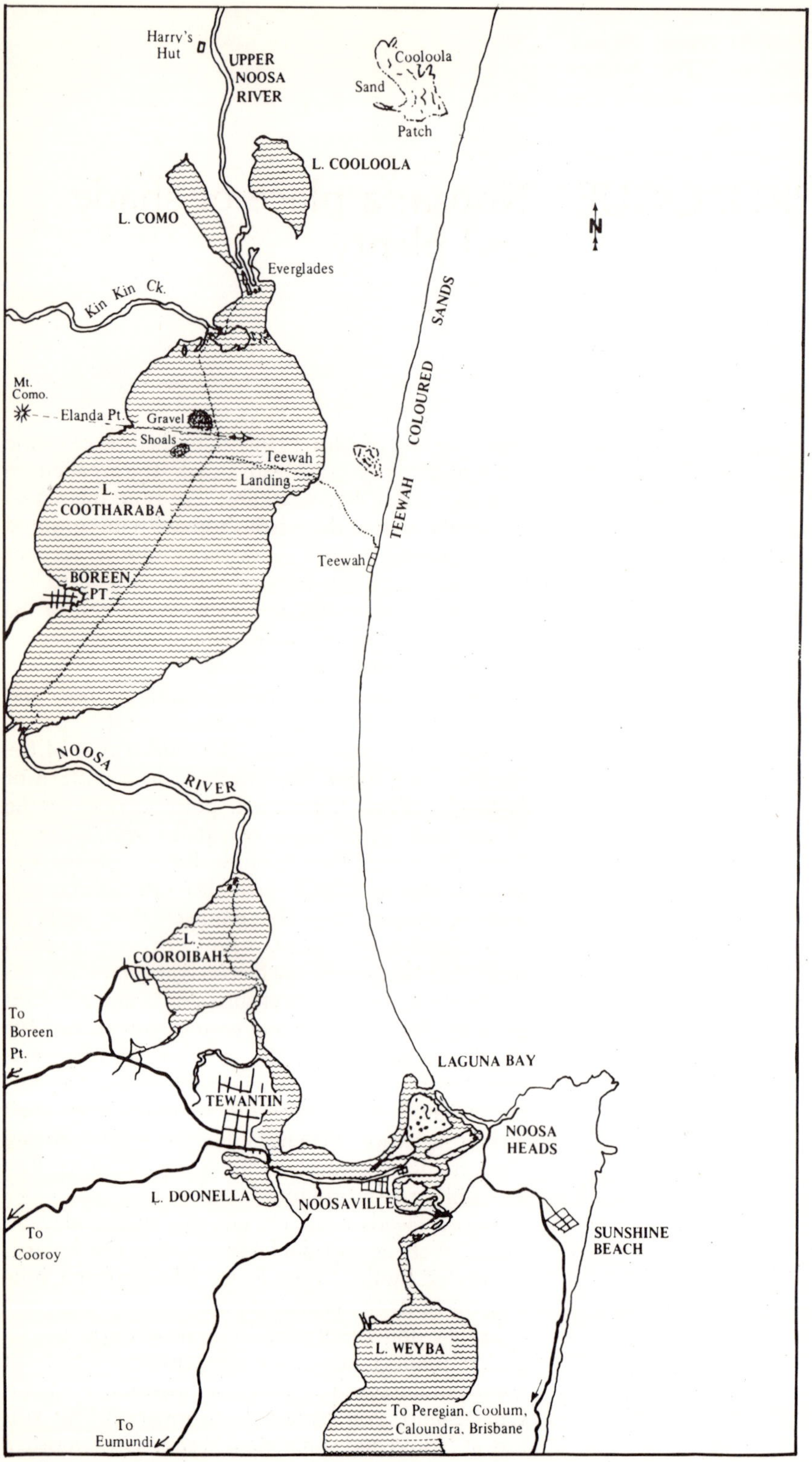

The Noosa River and lakes district. (By courtesy the Australian Coast Guard.)

2

Strictly, 'Noosa' refers to the whole of Noosa Shire, from Peregian to Pomona, from Noosa Heads to Tewantin and Cooroy. The Noosa estuary, the chain of brackish lakes, the freshwater upper reaches of the river, the national park, the long golden beach south of the Heads, the North Shore and the Teewah coloured sands, all are covered by that slightly exotic-sounding name.

It could be Red Indian; it is in fact Aboriginal, one of the few relics, apart from the scars on old trees where bark was removed long ago, of the vanished Dalla tribesmen of Noosa, and the Kabi-Kabi to the north.

Noosa is a place of ancient, unspoilt beauty and instant, man-made ugliness. The meaning of its name has been lost in the mists of prehistory, though it is believed to mean shade, or shady place. This could come from the cool rainforest below the headland, or the dark, dense Noosa Woods (now sadly depleted), or from 'a place of shades', for this was also a spirit place.

There is a certain high hill behind the national park with a circle of stones imbedded in the earth at the very summit. Outside the ring, saplings of new growth lean and posture like naked, dancing men. Inside that charmed circle nothing grows, not a blade of grass, a weed, a wildflower. It is not a typical *bora* ring (an initiation site). Was it the burial place of some dreamtime hero? And who, or what, keeps the circle free of growth? It is a mystery.

According to F. J. Watson, 'Noosa' is a corruption of *noothera* or *gnuthuru* ('shade' or 'shadow') — the coastal tribes had no sibilants.* The Dalla, who owned all the country with its river estuaries from Redcliffe, near Brisbane, north to the Noosa River, and west to Nambour and Cooroy, had a singularly euphonious language; it was a sub-dialect of Kabi-Kabi, spoken from Tewantin up through the lakes to Gympie and Maryborough on the Mary or Numaboola River. Both belonged to the Waka-Kabic group, all collected under the loose title of Dippil.†

Tree climbing with tomahawk and vine, circa 1865–70. Photograph from the collection of the late Mrs Bancroft. (By courtesy Caboolture Historical Society.)

The Dalla, the Kabi-Kabi, the Waka-Waka to the west who owned the bunya pine country, and the fierce Batjala ('Sea-people') of Fraser Island, once three thousand strong — all are gone, vanished from their former haunts.‡ In 1977, when a film was made about Mrs Fraser on the island, native actors had to be brought from Mornington Island in the far north of Australia.

For thousands of years they had enjoyed plenty, if not peace: oysters and crabs from the estuary, fish from the lakes, eugaries from the beaches, ducks and swans from the freshwater lagoons, brush turkeys and kangaroo, emu and wallaby

---

*F. J. Watson, 'Vocabulary of Four Representative Tribes of South-east Queensland; Also a List of Aboriginal Names and Their Derivation', 1944. Published as a supplement to the *Journal of the Royal Geographical Society of Australasia*, Volume 48, No. 34 (Brisbane, 1943–44).

†Dennis Bannister, linguist, expert in Aboriginal languages, in the *Courier Mail*, 1977.

‡Norman B. Tindale, 'Distribution of Australian Aboriginal Tribes: A Field Survey', *Proceedings of the Royal Geographical Society of South Australia*, 1949.

from the Noosa plain. They had heard, too, by the 'bush telegraph' of the murder by arsenic of over forty blacks at Kilcoy. This is believed to have been one of the reasons for the murder of the luckless Captain Fraser.

The Aborigines have left a legacy of place names along the coast — even Tin Can Bay, which was originally *tinthin*, 'mangroves'. The many words with double 'oo' sounds, like Cooloothin, Mooloolaba, Cooran, included one that was to become famous in conservation circles — Cooloola. (The *kululu* or *cooloola* is the coastal cypress, or callitris.) *Maroochy* meant 'red-beak' — the name for the black swan. Noosa Headland was called *Wantima*, 'rising up'.

The last of the local Aborigines, 'King' Tommy of Noosa, 'King' Brown of Woolumbah, and Susie, were rounded up by the police at Tewantin and taken away within living memory. They were getting old, and were taken to Murgon or Cherbourg 'for their own protection'. There were no children, and no young women. The south-east coastal tribes became extinct.

'The last of the tribe.' This picture adorned the bar of Tait's Royal Mail Hotel, Tewantin, for many years. 'King' Brown and Susie in cast-off clothing. (By courtesy Griffiths Studio, Tewantin.)

Poison, the gun, and the notorious black police had helped them on their way. So had an addiction to rum, which was used to pay them for their services in spearing fish and collecting new bark for roofing buildings. Many were shot under the official Queensland government policy of 'dispersal'.

Lieutenant James O'Connell Bligh had married his cousin Elizabeth, daughter of Captain Bligh of *Bounty* fame. He was appointed officer-in-charge of Native Police in the Wide Bay district in 1853. With his party of native policemen, recruited from Victoria, he blazed a trail from Traveston to Lake Cooroibah on the Noosa River, where they crossed over to disperse some blacks who had been spearing cattle. They shot down the blacks on Teewah Beach, where they had probably gone to collect pippies. The coloured sands that tourists visit today ran red with blood.

Nearer to Noosa Heads is Murdering Creek, which flows into Lake Weyba. Here a hunting party of seven white men, having sent one man on as a decoy, ambushed the Kabi-Kabi in their canoes. 'I never heard how many were killed on that occasion,' said the narrator, Mr David W. Bull, a pioneer settler of the area, 'but it is safe to say they were legion.'

A resident of Tewantin, Mrs George Clifford, recounted: 'My mother said that her mother, Mrs Elizabeth Hall, told her that in the early days, if the blacks were cheeky, the white men would clean their guns and go out hunting. Those "cheeky fellows" would never be seen again.'

An article in the *Queenslander* in 1877 said:

When the Blacks . . . have speared white men's stock, the native police have been sent to disperse them. What 'disperse' means is well known. The word has been adopted into bush slang as a euphemism for wholesale massacre . . . .

The savages, hunted from the places where they had been accustomed to find food, shot like wild dogs on sight, retaliate when and how they can . . . . Murder and counter-murder, outrage repaid by violence, so the dreary tale continues till at last the Blacks, starved, cowed and broken-hearted, their numbers thinned, their resistance overcome, submit to their fate; and disease and liquor finish the work we pay for our native police to begin.

Teewah, where the native police shot down the Kabi-Kabis, comes from *dauwah*, meaning 'dead wood' or 'dry timber', perhaps from a sand-blow there. The letters *d* and *t* are interchangeable, the spelling depending on the hearing and interpretation of the white man. The sound is somewhere between the two. Thus *Dauwadhun*, 'place of dead logs', became Tewantin — named after the sawmilling operations began in the 1860s. It was known first as 'the Short Cut', being the fastest route to the Gympie goldfields, which opened in 1867. (It was *not* called 'Tea-want-em' after a demand by the Aborigines for rations, as a persistent local legend has it, nor did Noosa come from 'No, sir!'.)

The bark canoes of the vanished tribes are now almost their only memorial. Not that any were preserved, but some of the giant old trees from which the Aborigines cut the bark are still standing. Right beside the main Tewantin–Cooroy road over the range is a canoe tree with the oval scar where a canoe was removed perhaps two hundred years ago with a stone axe. If you take the back track from Lake Cootharaba to Cooroy, you will come to a flood plain not far from the lake and here a group of big trees remain. Most have the scar of canoe or shield or *coolamon* (a shallow dish), with the bark welling back over the wound. A study of photographs of one old gum-tree taken at an interval of nearly one hundred years in South Australia shows not more than two centimetres of regrowth.*

When all about the lakes was a jungle of paperbark, mangrove and 'vine-scrub' (the laconic Queensland understatement for lush, sub-tropical rainforest studded with giant trees), the best way to travel was by water.

The winding waterways through the mangrove islands teemed with fish and mud crabs, while the riverbed was paved with mud oysters. The main street at Tewantin, originally called Goolloi Street (*goolloi* meant 'codfish'), was built from the huge Aboriginal midden on the river bank near where the Fish Board jetty is now. There were two heaps of shells, each mound three metres high. The Aborigines would also spear fish in the river and lakes, and cross to the North Shore to feast on *worwongs* — the large juicy cockles that further south were called *eugarie* and in New South Wales *pippi*.

It was a richly abundant land. Fish and shellfish, water birds and birds' eggs were there for the taking. No wonder the local tribes were fat and warlike, with such a heritage to defend. But after the massacres at Teewah Beach and Murdering Creek they quietly yielded their territory, and ceased to spear the invader's cattle when they were hungry.

---

*Robert Edwards, *Aboriginal Bark Canoes* (South Australia Museum, 1970).

One of the several 'canoe trees' in the Tewantin-Boreen-Point-Cootharaba area. This one is right beside the road on the Tewantin-Cooroy highway. The regrowth pattern indicates that the scar is about 200 years old. (By courtesy Kathleen McArthur.)

Of the original inhabitants of Laguna Bay, with its curving shore and cliffs of coloured sand, its freshwater springs bubbling up through the beach, not one is left. Today, monstrous 'vacuum cleaners' suck up water and sand and spew it out in dirty heaps, extracting rutile. Plastic bags, beer cans, milk cartons, old tyres, rubber thongs and tangled nylon lines are mixed with the seaweed at high-tide mark. The black men left only their heaps of empty shells.

Now the trees, which gave Noosa Heads its Aboriginal name, are fast disappearing. The fragile, sandy hills are eroding; the Noosa Woods are going; the lines of ancient she-oaks that lined the foreshore have long disappeared. Today, a rock groyne and a bare, pumped area of sand have replaced the golden, sculptured sandhills where rainbow lorikeets fed on the nectar of the coast banksia and azure kingfishers dug their tunnel nests. Swimming pools, neat green lawns, modern motels and units line 'the esplanade'; and the birds, like the blacks, have departed.

Where the estuary used to wind in turquoise loops among spits of clean, yellow sand, past the dark cluster of Noosa Woods and among green, wooded islands threaded by narrow channels and fringed with mangroves, there is now a canal development, right in what was once the mouth of the Noosa River.

The sandflats where fish used to breed, the mangrove flats where mud crabs abounded, the tall kauris and blue gums where sea eagles had their nests have all been removed, or built up with a cubic metre's depth of sand covered with 'instant grass'. Houses are being built in what was part of the river's natural flood plain, and the sandy channels are now edged with rocks.

Yet only ten years ago the estuary remained almost in its pristine state. Apart from a few fishing dinghies, a passing trawler, the white sail of a yacht against the brilliant water, it looked from the hill above much as it had on the day the first white men stood here and surveyed it.

That was a hundred and forty years ago.

1 'A picturesque beach'

Mr Andrew Petrie and his companions . . . in May this year carried out an expedition to the near North Coast of vital importance to the Colony. Mr Petrie's object was to extend his knowledge of timber resources, and the others were anxious to find good sheep land . . . . After a day's sailing, the party landed with difficulty at a picturesque beach where a tribe of natives had gathered.

Queensland Gazette, 1842

When Captain Cook first sailed this coastline, he stayed well out to sea. He named Double Island Point, which was not an island, and mistook Fraser Island for part of the mainland. He did not see the mouth of the Noosa River at all. When he finally came inshore and landed at Bustard Bay he formed a poor impression of this part of Queensland. He was bitten by green tree ants, stung by stinging caterpillars, harried by clouds of mosquitos; he got grass seeds in his trousers, and there was no fresh water. This was in May 1770.

Matthew Flinders, in the *Investigator* voyage of 1801, passed Noosa Headland at about seven in the morning. Double Island Point was nearly fifty kilometres to the north-west:

. . . and the shore abreast, a beach with sandy hills behind it, was distant six miles . . . . Between the point S.W. and a low blue head was a bight in the coast, where the sandhills seemed to terminate . . . . At 9.30 hauled around Double Island Point, where a number of Indians, fifty or so, were looking at the ship . . . . Saw their fires on shore at night.

Today, the look-out on a ship sailing northward along this coast would still see flickering fires on the beaches and headlands, but he would not see any of the 'Indians' after whom Cook named Indian Head on Fraser Island. The modern nomads are car-borne, travelling up the beach in buggies and four-wheel-drive vehicles, camping out and cooking their day's catch over open fires. The *worwongs*, or eugaries, are still as thick as pebbles on the beach, though the black men who used to eat them, leaving whitened heaps of their empty shells in sheltered hollows, are now no more. Only the oystercatchers eat them today, opening them with their red bills, and the white men gather them for bait.

It was not the fishing, the view, the surf, nor the climate that brought the first settlers to this area, but the big timber of the hinterland with its high rainfall and rich soil. This it was that spelled the doom of the Noosa tribes.

Four explorers from Brisbane Town were the first free white men to see Noosa Heads beach: Messrs Andrew Petrie, Henry Stuart Russell, Wrottesley and Jolliffe. With a crew of convicts and two Aborigines, they set off in May 1842 for the 'Marouchy River', which was marked on their charts. Petrie had heard that there were stands of valuable timber just inland from the north coast.

They made 'Madjumba' Island (Old Woman Island off Maroochydore) the first day, then set sail northward and arrived shortly after sunset in Laguna Bay:

This bay or inlet has a river in the bight, which forms several large lakes, or sheets of water. A few miles inland from one of these, Mrs Fraser was rescued from the Blacks by Bracefell.*

(Bracefell, or Bracefield, was a convict who had escaped from the notorious Moreton Bay Penal Colony years before, and lived with the tribes as one of them. He dreaded the lash too much to return with the rescue party of the shipwrecked Mrs Fraser in 1836.)

As they arrived, twenty or thirty blacks, unarmed, appeared from the low brush bordering the beach; as there was too much swell running for the boat to be beached, they waded out and carried the white men ashore in return for some biscuit. They seemed 'bold and daring', and Petrie kept his loaded rifle pointed at them while they also carried the luggage ashore.

The next day they sent a message to Bracefell, who arrived with his tribal 'father'. He wept openly. He knew his own name, but could speak little English, though he knew the dialects of four different tribes. He was known to the Aborigines as Wandi.

Stuart Russell wrote:

At this time, little was known of the coast here; no close survey had been made except from seawards . . . . The low bluff which formed the southern and most eastern point of the bay in which we were, Petrie now called 'Bracefell's Head' [Noosa Head] . . . . From a higher ground further back we could see several noteworthy eminences, and of these Bracefell and his friend told us the native names, which were written down on the spot; Mandan, Cooroora, Coollum, Cooroy, Yura, Eirange,

and Boppol . . . . Signal fires were rising rapidly in every direction from native camps, telling the news of our arrival. [The north shore] . . . appeared but solitude and desolation, with steep sandhills.†

Later they walked with Bracefell to 'a corroboree ring scooped out of the earth in the fashion of an immense circus', the earth around it forming a low mound. But neither Wandi nor Davis (Durrumboi), whom they rescued at Wide Bay, would ever tell the details of the rites of 'man-making' ceremonies in which the ring was used as a theatre.

From the Aborigines at Noosa Heads and at Wide Bay, Petrie must have learned of the great stands of timber beyond Lake Cootharaba and Kin Kin Creek. Giant kauri, hoop pine, blackbutt more than thirty metres high, red cedar, ironbark, blue gum, rosewood, forest oak and tulip wood, the *myrara* with huge spurs supporting the trunk, all were bound together by trailing vines, lawyer canes, flowering creepers, ferns and elkhorns to make a 'vine scrub'.

An anonymous correspondent of the *Wide Bay and Burnett Times* of Maryborough, thought to be none other than Lieutenant Bligh (who had taken up some 6500 hectares with a frontage to Lake Cootharaba in 1860, before survey), wrote:

The principal timber is kauri, of large growth, and it stands thicker on the ground than any scrub I have seen on the Mary . . . . This locality opens an inviting field; and from the beauty and diversity of the surrounding scenery would form a delightful locality for a settlement, within easy distance of the metropolis.

Hardy timber-getters and their families were soon attracted to the area. The valuable red cedar is now almost extinct. There was so much of it in the early days that it was used for house fires and for chook-houses. Sometimes trees were felled that were just too big to cart away, and they were left on the ground to rot.

The rush to get the valuable timbers started in 1865, before even the 1867 gold rush to Gympie. Development followed the usual pattern: first the explorers and geologists; then the exploiters, tearing everything valuable out of the soil, denuding the hillsides of forests, destroying wildlife, and 'dispersing' the Aborigines, the land's original inhabitants, when they objected to losing their hunting grounds. Then came the farmer-settlers, clearing with cross-cut saw and fire, bringing in cattle and dairy herds to munch the wiry grasses and muddy the creek banks. Koalas, once prolific in the gum forests and often seen about Tewantin and Noosa Heads, lost their habitat. Those that

---

Tom Petrie's Reminiscences of Early Queensland, edited by Constance Campbell Petrie (Brisbane, 1904).

†H. Stuart Russell, *The Genesis of Queensland* (Sydney: Turner and Henderson, 1888).

Hay's Bayview stands on the skyline, overlooking the paperbark lagoon. The lagoon is now the site of Ocean Breeze units, a large new development. This picture was taken around 1927. (By courtesy Olive Freeman.)

survived were shot out in the great slaughter of 1927, when an open season was declared on their 'valuable' skins. These harmless, infinitely attractive small animals are now rare in the local bush.

At Noosa Heads, the high, rolling sandhills were once covered in a stand of large cypress pine. Where they sloped towards the south and were open to the south-easterly winds, a wild jumble of banksia scrub was inhabited by spiny anteaters, bandicoots, dingoes, tree snakes, carpet snakes, and a million honey-eating birds. Today, these sandhills are covered in holiday houses, permanent homes and multistorey units.

If you want to catch a glimpse of how it was, walk up the public footpath behind the site of the old post office below Halse Lodge. Two turns of the track and you are in dense bush, almost jungle, with tree orchids and flowering creepers. A little creek tumbles over a waterfall at the bottom of the Marriott-Burtons' private one and a half hectare garden. From the top you can walk on up the road to Laguna Lookout, passing from native pine (callitris) to hoop pine forest, and then to

Transport on foot: boats used to anchor at Noosa Woods after entering over the river bar. From here guests would walk through the sandhills to Laguna House. Here stores and fish are being transported to Hastings Street. (By courtesy Olive Freeman.)

A very wet season in the 1920s. The three blue gums are noticeable on the left.

The Miss Laguna II *at her moorings in the lagoon, inside the bar at Noosa Woods, circa 1948. The main channel of the river ran through here until recently. (By courtesy Kevin Freeman.)*

open eucalyptus scrub on the stony soil towards the summit. Below the lookout, the rich, red clay supports a banana farm, the last relic of Freemans' Noosa Vale (now sold to developers).

In front spreads the view seen by Petrie in 1842, Lake Cootharaba in the distance and blue volcanic peaks dotting the horizon. The surf comes round the headland in its age-old pattern, but the looping river, the patterns of sandy or vegetated islands, paperbark lagoons and densely wooded, natural beach have all been changed by the hand of man. On the present site of Ocean Breeze units was a large, shallow lagoon surrounded by paperbarks, the haunt of water birds and native bees. The sand-blow opposite Laguna House, where generations of guests walked through to the beach, is filled with yet another shop and home unit complex, four storeys high (Fairshores).

Continuous motor traffic sweeps up and down Motel Hill and in holiday times clogs the dead-end Hastings Street at the bottom. One of the last areas of untouched tidal flats, at the entrance to Noosa Sound, has been 'reclaimed' for a car parking area. And the whistling kites whose ancestors were nesting in the tall blue gums opposite, long before the first explorers came, have to fly further and further from home to feed their young. The 'picturesque beach' has become a car park with sand and sea bathing attached.

2 Almost an island

One could see much evidence for despair at what we have done in the last twenty-five years — for the errors, for the lack of foresight, and for our inability to predict what was likely to happen and to take steps quickly to prevent it. Coastal development has so far been done on the cheap.

Conservation of the Australian Coast, Special Publication No. 7 (Australian Conservation Foundation, 1972)

If you cut the small detached masses of Stradbroke and Moreton Islands from the map of south-east Queensland, and place them over the Sunshine Coast north and south of the Noosa estuary, you will find that they fit almost exactly into the spaces between Double Island Point to the north and Maroochydore in the south.

The coast proceeds northwards from Brisbane in a series of sandy scallops. After Stradbroke and Moreton comes Bribie — scarcely an island now, as a bridge joins it to the mainland — then the Caloundra–Mooloolaba coast, the Noosa sand mass, and the Cooloola sand mass north and east of the Noosa River. Then comes the great detached mass of Fraser, or Great Sandy Island. Each is roughly wedge-shaped, with the wider and higher end to the north; some of the dunes reach over two hundred metres in height.

How were these huge sand masses created? All had the same origin: sand washed down from the mountains by the northern rivers of New South Wales was gradually carried hundreds of kilometres northward by the littoral current. Deposited wherever some rocky obstruction slowed the current, the sand became joined into a series of huge sandspits formed by the prevailing south-easterly winds and waves. Stabilized by vegetation taking root along the foreshore as they reached above the highest tides, the gradually widening dunes were able to provide shelter for a secondary row of plants to gain a foothold until, in time, the moving dune was tamed, and a new wind-oriented, wind-shaped but stable landform was created. A rising coastline pushed the sea back, but in geologically recent times the Noosa and Cooloola masses were islands, while the beaches were formed no more than five or six thousand years ago.

A chain of old, dry lake beds and swamps (and

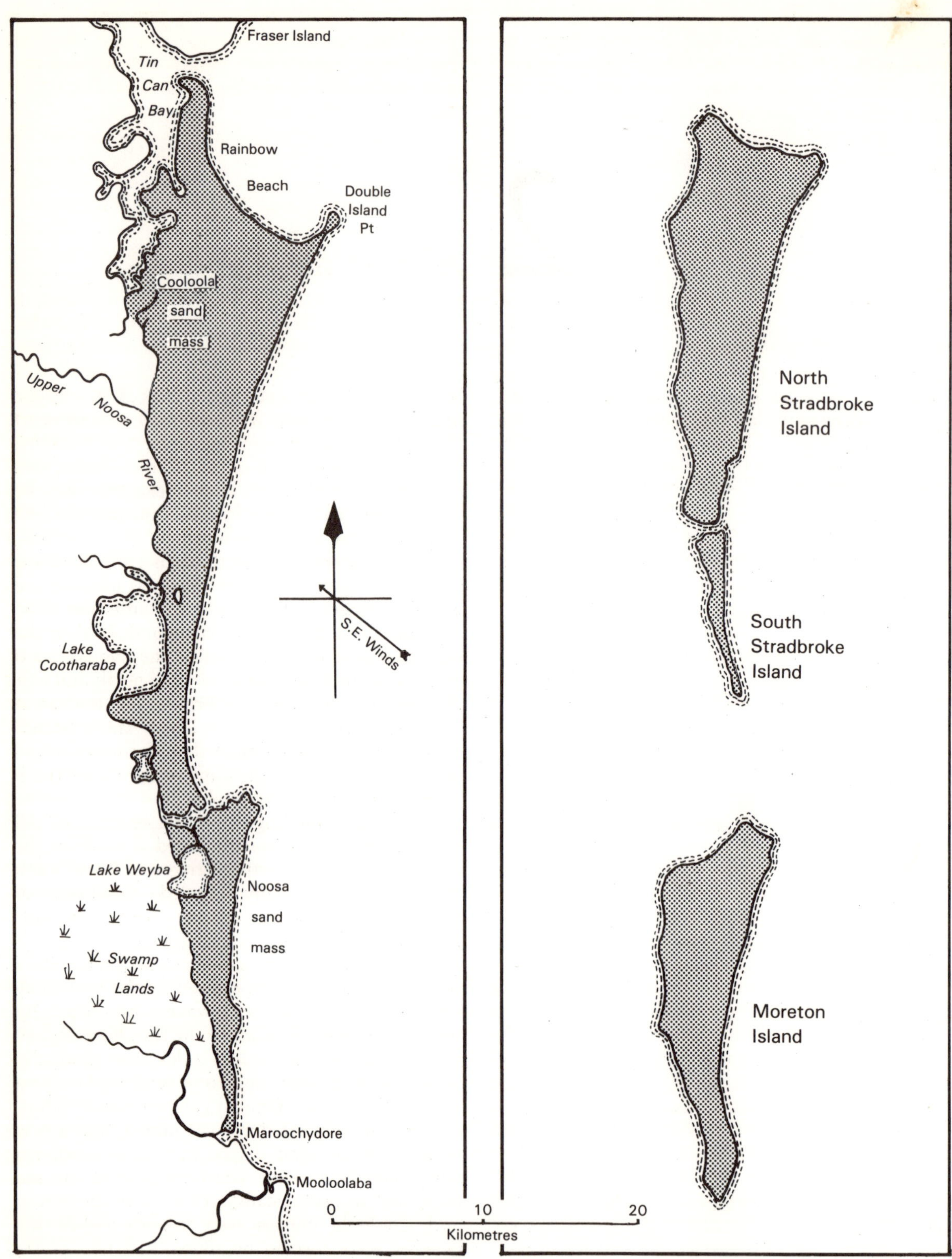

Relative shapes of the Cooloola sand mass, Noosa sand mass, and the islands of Moreton Bay.

one shallow salt lake and creek, Weyba) shows where the sea once ran between, just as Pumicestone Passage runs behind Bribie today. The Cooloola mass is more nearly an island, for the north–south channel of the Noosa River runs through a chain of brackish lakes that rise and fall with the tides, while the deep inlet of Tin Can Bay to the north almost reaches the fresh headwaters of the Noosa at Teewah Creek (see map on page 12).

Because of this skein of waterways and the expanse of barren heath and paperbark swamp to the south, the rocky headland and sandy mouth of the Noosa River (where Noosa Heads township stands today) was almost as cut off as an island in the ocean. The only access was by sea. Thus Tewantin, on a better and deeper anchorage further upstream, and connected by road with Gympie and Maryborough, was developed first.

It was this bastion of rock in the sea of sand, cutting off the prevailing south-easterlies and the cold southerlies, that later led to the development of Noosa Heads as a leading beach resort. Yet there are people alive today (for the marvellous climate tends to longevity) who can remember when there were only four dwellings between Weyba Creek and the Noosa Woods.

Even by the 1920s, long before Noosa was discovered by Melbourne and Sydney people, Queenslanders in the know used to flock there by train and boat to enjoy the warmth, the sparkling sea, the wide beach of yellow sand, the welcoming shade of huge, spreading trees, and the hospitality of Hay's Bayview and Donovan's Laguna House.

On the south side of the headland, exposed to driving, salt-laden winds, there evolved a heathy flora of stunted banksia, wedding bush, casuarina and grass-tree, bracken and blady-grass. These bind the dry, white, delicate sandy soil to make a thin cover of humus. A view taken from the air only fifty years ago would reveal a dense mass of foliage with only here and there a white track showing like a parting in dark hair.

To cross the headland to the sheltered north-eastern side is to enter a microclimate that is almost tropical. It is a different world. The stunted, writhen, ground-hugging heathlands turn to dense, green rainforest. In the north-eastern corner of the headland, where the 1800 millimetre rainfall comes channelling down the rocks, and the northern sun is trapped every afternoon, there is even a pocket of towering scrub with jungle trees, palms, vines and ferns.

Further up the hill grow giant blue gums, hoop pine, silverleaf ash, blueberry ash and lilly-pilly.

Down the slope leading to the sandy spit that is now Hastings Street, a track was cut, through thick bush twined with purple convolvulus and sarsparilla, past a paperbark swamp and lagoon to the wide, sandy beach.

A unique flora flourished on the huge masses of pure sand, up to 150 metres high, that extend inland from the headland in waves. The soil lacks practically everything in the way of minerals that garden plants require and yet here you will find old-man banksias of huge growth, scribbly gums, their smooth trunks tattooed with dark 'writing', bloodwoods breaking into creamy blossom in summer, and the brilliant green of the native cypress or Cooloola pine. Here the wedding bush (*Ricinocarpus pinifolius*) becomes a small tree, covered with starry white blossom in spring; and the shy green-winged pigeon blends with the green and brown of lions' tails and bracken. On the low-lying ground, blossoming paperbarks (melaleucas) provide the best nectar for honeybees, and the long-leafed Moreton Bay ash is a gum that has adapted itself to withstand all but the fiercest salt winds. The whole area is one of extraordinary diversity.*

This green 'island' is enclosed on all sides either by the sea or by the 'everlasting wallum', which stretches from Caloundra to Tin Can Bay. The coastal wallum was hated almost as much by the first settlers as the 'everlasting mallee' was in South Australia and Victoria — 'the miserable Wallum Country, where animal and vegetable life seem equally absent; the trees seem to shrivel into shrubs; grass is invisible; and bird and beast are alike conspicuous by their absence', as the *Maryborough Chronicle* described it in 1890.

In fact these poor, acid, sandy soils grow a profusion of wildflowers and support colonies of

---

*Botanists at first were puzzled by the diversity and density of the coastal scrubs, growing in almost pure silica sand that was deficient in all mineral nutrients when the masses were formed. According to Dr. L. J. Webb of the C.S.I.R.O.'s forestry branch, the first nutrients were deposited thousands of years ago in a thin layer by sea-spray blown by the wind and evaporated on the sand. Salt-resistant small plants could then develop from seeds dropped by birds or carried by winds. As each plant and rootlet died, the nutrients it held were returned to the sand and picked up by neighbouring plants. Thus increasingly larger plants, and eventually trees like the large kauri and cypress pines found on Hay's Island, could be supported.

Ecologically it is known as a 'closed cycle'. However, once man interferes the cycle is broken, for trees are cleared and carried away, or burnt so that their ashes are leached away by rain; and the nutrients that may have taken thousands of years to accumulate are lost.

honey-eating birds, besides the migrant flocks of rainbow lorikeets seeking banksia nectar in season, and black cockatoos after the banksia seeds from the old-man cones. For many years this wilderness of swamp and heathland, fronted by long stretches of beach open to the full force of the Pacific Ocean, was seen as completely useless. A good idea of its extent can be gained from the top of Peregian Hill, a rocky volcanic outcrop near the coast.

A few tried running cattle there on 'permissive occupancy' leases — Frank Crank was one — in a fenceless area where the cattle wandered down to the beach and browsed on salty grasses that gave them needed minerals. But 'thieves and ticks' made the area unworkable after about 1920, Crank says. Then came World War II, and someone had the bright idea of using the land to fire practice shells over, from the high rocky point at the top end of Sunshine Beach. Live shells were used; after all, the country was worthless and only a few kangaroos and emus might get killed. To this day, the area west of the coastal highway, and south from about opposite Castaways, is dangerous to cultivate. One holding, leased from the Lands Department, was taken back after a plough just missed a live shell. Two more shells were exploded near the road in 1977, Army experts having been called in to detonate them. The only warnings are two rusted notices posted on the highway:

NOTICE RE UNEXPLODED SHELLS

South to Lake Weyba and Murdering Creek on the west, from the coast on the east and north to a line to the coast where Weyba Creek enters the Lake, was used during the 1939 war and post war as an Artillery Range . . . .

Suspicious objects should not be interfered with but reported by the quickest means to the nearest police station . . . .

DO NOT HANDLE

Thirty years ago the first developer, Alfred Grant, saw the possibilities of Noosa Heads, and subdivided the hills behind the bay that face west and slope away to the south, giving magnificent views of the Noosa River estuary, Weyba Lake, and the coast south of the headland. The roads were impossibly steep and badly engineered, and, where denuded of growth, the fragile sandy soils began to slip downhill with the heavy summer rains.

Most of Noosa Heads township, apart from 'Snob's Hill', the exclusive rocky headland leading to the national park and Little Cove, is built on filled-in swamps or these gigantic sandhills. They form a natural amphitheatre where houses and motels have been built in tiers, with sweeping views of mountains, river, lakes and sea.

The shire council had to restructure and seal the roads, and build covered stormwater drains to take the run-off from the steep hills. Gradually the blocks were sold, though few were built on, until at the end of the 1960s and early 1970s there was a boom: retired families and spec. builders moved in, and soon the bushland greenery was giving way to brick, block and chamferboard. The tiny, twenty-perch blocks of the original survey were retained, leaving no room on a single block to keep any bushland on either side. Houses are now cheek-by-jowl, each with its piece of neat lawn in front and the obligatory umbrella trees. Suburbia has come to the country.

3 The first land speculators

In those days the channel for boats was much nearer the south Head . . . . From the deck [of the paddle-boat *Culgoa*] we could see the people on the balcony of Bayview, the home of Mr Walter Hay. He had a fine orchard to supply his boarders with fruit. From his boarding house a splendid view of all the bay and of the inland country could be obtained.

Reminiscences of Mrs S. K. M. Hartley, 1890s

Men who had been drawn to the unsettled area between Brisbane and the busy northern sugar port of Maryborough, by the discovery of gold at Gympie and of the valuable stands of timber in the Kin Kin scrubs, often stayed to take up land. They became cattle breeders, butchers, farmers and fishermen. Some opened sly-grog shanties for the thirsty sawyers and bullock drovers, and many a lonely farm had its own still producing illegal whisky.

The first settlers had to clear jungle and rain-forest swarming with snakes, stinging trees, stinging caterpillars and sharp-clawed lawyer canes. They had to drain paperbark swamps alive with leeches and mosquitos and sandflies, and make their own roads, bridges and wells.

The first-comers took up the best river-front and lake-front land, near enough for access by water but high enough to escape flooding in normal years. The first settlements were on Lakes Cootharaba and Doonella, and at Cooloothin Creek. Elanda Point, or Mill Point, was at the top of Cootharaba, while Tewantin grew up in the corner of Lake Doonella and the upper Noosa estuary.

As early as 1866, timber-getters had explored the back country of what is now called the Sunshine Coast. The rivers made the first highways through the thick scrub.

In about 1869, Messrs McGhie, Luya and Goodchap developed the industry in the area. A timber mill was set up at Elanda Point, the nearest point on the lakes to the Kin Kin scrubs, for the simplest method of transport for the sawn logs was by water. They were brought to the mill on sleds drawn by draught horses or bullock teams, and later on a tramway with horse-drawn trucks.

By 1870, a complete township had grown up in the midst of the unpeopled bush. Tewantin

Joseph Keyser, Tewantin's first fisherman, after whom Keyser Island is named.

became its port, Noosa Heads its front gate. Nothing of it now remains but a few lonely graves — their iron headstones tumbled down — the old tramline, and the stumps of the wharf.

Supply boats came each week from Brisbane and Maryborough. Cobb and Co. had started a coach run from Brisbane to Gympie in 1868, but it was slow and hazardous. For years all supplies for Elanda Point and Gympie came over the Tewantin wharf, known as 'the Short Cut', by way of the Noosa River bar. Passengers, too, preferred the comfortable steamer passage as far as Tewantin (fare twenty-five shillings for cabin passengers, later reduced to fifteen shillings) to being bumped and bruised for three days in the coach from Brisbane. Coaches for Gympie would meet the steamer at Tewantin.

The Noosa estuary was first surveyed by Commander G. P. Heath, Port Master for Queensland, in the tug *Brisbane* in 1869. He found a depth of around two and a half metres over the bar at high tide, and three to four metres in the lower river. His plan, issued in 1870, referred to 'Nusa Harbour'. 'Neusa' was another contemporary spelling.

The first paddle-steamer to dodge in and out over a notoriously tricky bar was the stern-wheeler *Gneering*, belonging to William Pettigrew, which took unsawn timber down to Brisbane. After McGhie/Luya's big timber mill was built at Elanda or Mill Point, a new steamer,

First road to Cooroy from Tewantin: on the way to the Gympie goldfields.

Bullock wagons hauling timber from the big scrubs, Cootharaba area.

Above: *Rafting logs, Kin Kin Creek, circa 1910. (By courtesy Oxley Memorial Library, Brisbane.)*

Below: *The paddle-steamers came early: the Adonis at Dath/Henderson's Wharf, Tewantin, 1880. (By courtesy Bill Griffiths.)*

Laguna Bay and 'Nusa Harbour', 1870. Traced from Captain Heath's plan. (By courtesy Department of Harbours and Marine, Brisbane.)

Above: *Dath/Henderson's sawmill at Tewantin, 1880. The ketch* Noosa *is at the wharf.*

Below: *Still going:* Noosa *supply ketch at Tewantin in 1921. (By courtesy Bill Griffiths.)*

The supply boat Goldcrest, on her weekly trip from Brisbane, aground on the North Shore at the bar. She has been tethered with a rope to a tree. By the next day, following a high tide, she had been broken up by the waves. (By courtesy Olive Freeman.)

the *Culgoa*, brought supplies to the sixty people, workmen and their families. She was a side-wheeler, twenty-seven metres long and drawing just under two metres. She could not go above Tewantin. The *Culgoa* could take 100 cubic metres of timber. She was later wrecked on the bar (1891). Between 1870 and 1892, small 'droghers' (paddle-steamers), the *Alabama*, the *Elandra* and the *Black Swan*, shuttled between Tewantin and the top lake, coming down with pontoons of timber in tow.

Other supply boats in the 1880s were the *Adonis* (side-wheeler) and the *Arakoon*. The auxiliary ketch *Noosa* followed and, as late as 1927, the *Goldcrest* was still coming on a weekly run with beer, kerosene and other essentials from Brisbane. She too came to grief on the bar and broke up on the north shore in 1928.

Some years after a tragic boiler explosion, the mill was abandoned and, in about 1900, Dath/Henderson started a new timber mill on the northern bank of the Noosa River opposite the present site of Tewantin. It was known then as 'Coolloy' from the Aboriginal word *cooloi*, meaning 'shark'. At this time there were more settlers on the far bank than on the south side, where the main part of the town is today.

Besides sawn timber, other exports to Brisbane were honey, fish, and oysters from the river and lake beds. Smoked fish was also taken overland to Gympie.

Up to a hundred years ago, there was very little marked on the government map of the 'Gympie District' apart from the Gympie goldfields, 'the port of Noosa', and some timber and rafting leases around Cooloothin Creek. In 1877 most of Noosa Heads, including the present national park and Sunshine Beach, was still marked 'Reserved for Aboriginal Use'. The decimated tribes had become fringe-dwellers about the small township of Tewantin, living on hand-outs of flour and tea, working desultorily as stockmen around Lake Weyba, or selling fish and stripped bark to the townsfolk in exchange for rum and tobacco.

After the survey of 1870, one of the first to select land on conditional purchase was Seaton H. Dun, who held Lease No. 111 on Cooloothin Creek, described as a 'rafting ground'. It was transferred in 1881 to A. F. Luya of the Good-chap/McGhie/Luya timber interests. Mr Grainger Ward's selection was the first in Tewantin proper, and was partly resumed later for the township reserve.

There was some argument about the boundaries of Ward's selection, No. 55, because he had selected before survey, in 1869. The pioneer bushman Walter Hay, a canny Scot who had already acquired more than 200 acres at Tiaro south of Maryborough, had also selected close by, taking some of the best high land on the river bank in what was to become the town of Tewantin, plus 120 acres of hinterland.

In 1870, Hay wrote indignantly from the Gympie Lands Office to the Surveyor-General in

Goat Island.
NOOSA RIVER.
M37.241
DOONNELLA LAKE.
Ford
Mangroves
106
F. G. Goodchap.
80 . 0 . 0
M37.35
(2544) 270°
2290
151
1652
167
W. Hay.
80 . 0 . 0
M37.93
90° 2474
90° 2000
Walter Hay
336 . 1 . 0 Ex Road.
M37.63 &W39.224
90° 2676
90° 2592
M37.322 &W39.224
270° 5025
266°20' 1350
243°18' 1418
True Bearing 8°52'.00"
M37.790
90° 8268
6951
10 v
John Monk
319 . 3
(10178)
9102.6
883.7
254°20' 390
280°22' 1770
R 2691
RESERVE
FOR
PUBLIC PURPOSES
Ab 57 . 0 . 0
Eenie Ck.
M37.798
3442 270° 5000
1558
1667
Joseph Keyser.
109 . 3 . 0
M37.772
270° 3996

P A R I S H

22

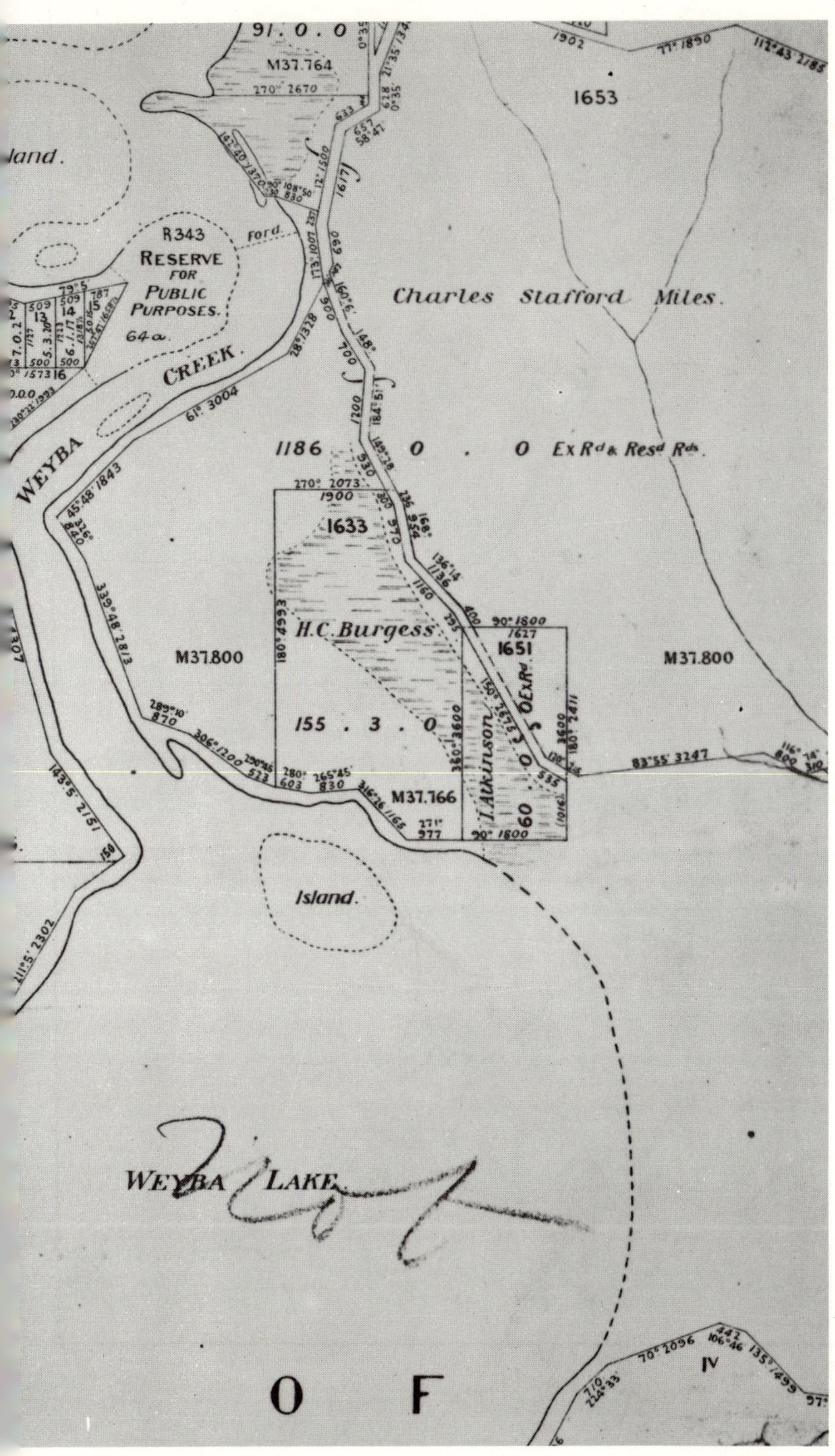

Parish of Tewantin 1896, showing fords over Doonella and Weyba and early landholdings in Noosa area. (By courtesy State Archives, Brisbane.)

23

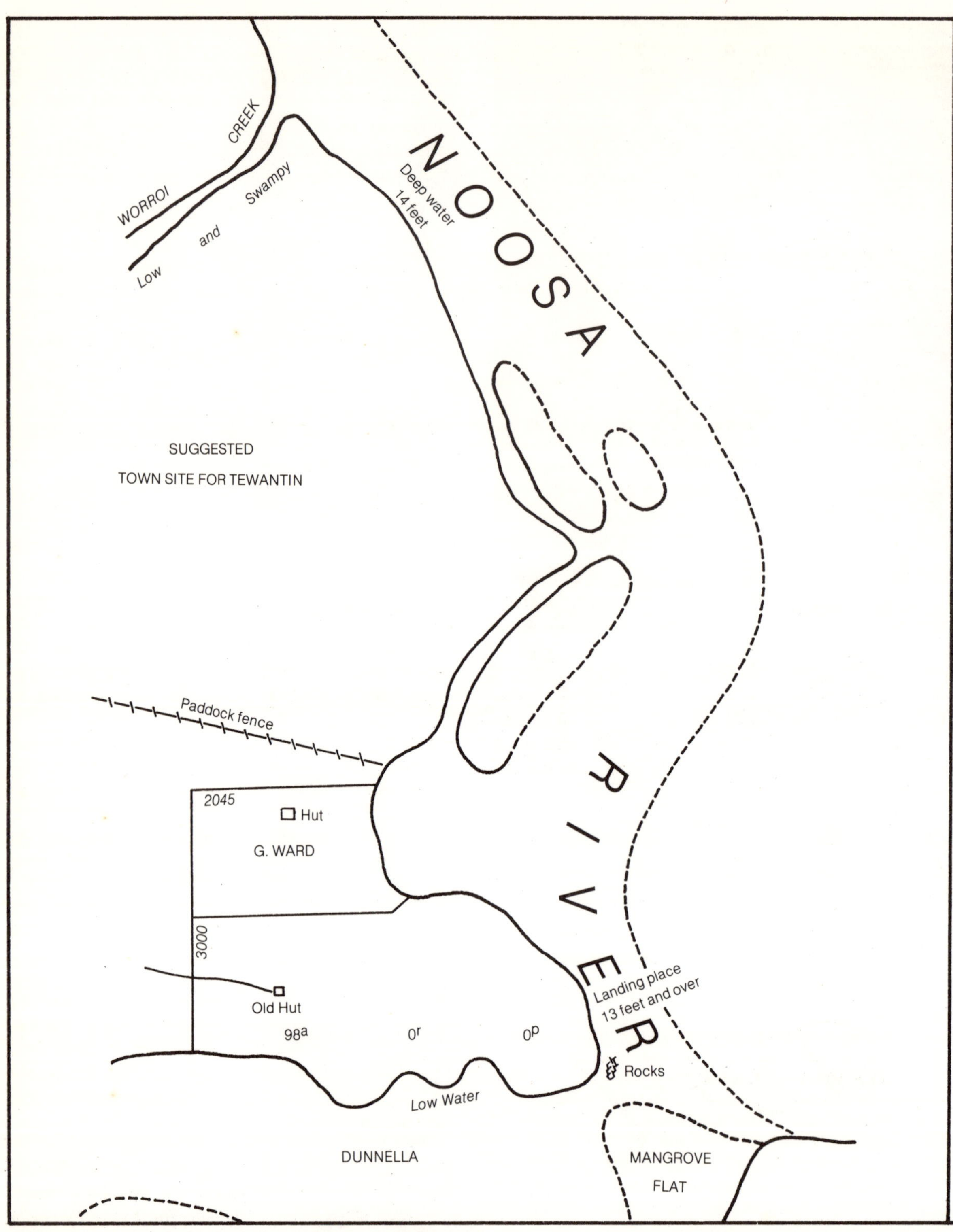

'Ward's Estate and suggested town site for Tewantin.'
From a plan by the Lieutenant Surveyor, Clarendon
Stuart, 1870. Note old spelling of 'Dunnella' Lake. (By
courtesy State Archives, Brisbane.)

Pioneer settler Walter Hay in 1905, aged about seventy. (By courtesy Robinson Studios, Nambour.)

Brisbane, protesting that the survey of 1870 was not accurate, in that his own selection No. 82 was now overlapped by Ward's, whose application had gone in six months earlier than his.

As it turned out, Hay's selection and most of Grainger Ward's were resumed for a township reserve. Ward was allowed to select an area to the north of the reserve (Ward's Estate) as compensation. Rent and survey fees were refunded to Walter Hay. His letter spoke of his pioneering trip from Gympie, which he felt Ward had taken unfair advantage of:

Since [applying for the land] I have been employed at considerable expense in marking of a road from here to Noosa. I was about seven weeks over it. After which I then proceeded to Rockhampton for the *Sir John Young* steamer of which I am the owner and piloted her up the 'Noosa River' where she now lies for the purpose of trading between there and Brisbane.

After arriving at Noosa I found that part of the peice I applied for had been surveyed into Mr G. Ward's peice for what reason I am at a loss to conceive, except that Mr Ward expects to reap the benefit of my Capittal and labour . . . .*

---

*Manuscript letter, Lands Department, Files 1868–96, Queensland State Archives, Brisbane.

Hay went on to say that there was nothing on the bank (north of Doonella Lake entrance) but a few slab huts left there by some timber-getters in 1866, and that Mr Ward himself told him the 'peice' adjoining was open for selection; he appended a sketch of the area, and concluded thus: 'I beg you will be so kind as to reply by Telegram or letter at your earliest convenience.'

In January 1871, after survey, F. G. Goodchap took up the first land south of the river and east of Lake Doonella. This 80 acres adjoined a mangrove swamp where the Lakes Motel stands today, and was a fording place for cattle and horses from Tewantin. Adjoining Goodchap's was the second large holding of Walter Hay, Lease No. 151 of nearly 350 acres with river frontage — 'second class pastoral land'. He called his home 'Hilton Park', which gave its name to the present Hilton Terrace, Noosaville. He had put in a crop of maize for cattle fodder as soon as the land was leased.

When Hay applied for his deed of grant, having complied with the residence clause and improvements clause of the Land Acts, he described the improvements as consisting of 'a cottage of three rooms with veranda, outhouse for servants, slaughterhouse, stock yards, garden and some clearing . . . . I was in continuous bona fide residence from July 1871 to 1873. Improvements valued at £169'. The freehold was granted in 1878.

Walter Hay had come from Maryborough (where he was a licensed waterman, working the ferry over the Mary River to Granville) by way of Gympie. He blazed a track through the bush from Gympie to Tewantin and set up his camp on the banks of Lake Doonella. A group of local Aborigines, always inclined to be 'cheeky' with a lone white man, came over in their canoes and demanded tucker. He gave them some, but when they asked for tobacco he refused their request. One of them threw a spear, but at that moment Hay had bent to look in the camp oven, and the spear hit him a glancing blow on the back. He had a loaded revolver handy and fired at their feet. They departed hastily; they already knew the power of guns. Hay was said to have borne a scar for the rest of his life.

Among other selections, Hay took up land on the south side of Lake Doonella, on the Noosa River just above Lake Cooroibah, and 204 acres of the best flood-free land on the west of Lake Weyba. In 1880, having lost the papers for his selection No. 115, which he wanted to transfer to W. L. Smith, butcher, he made a declaration giving his address as 'Hilton Park' (now Noosaville).

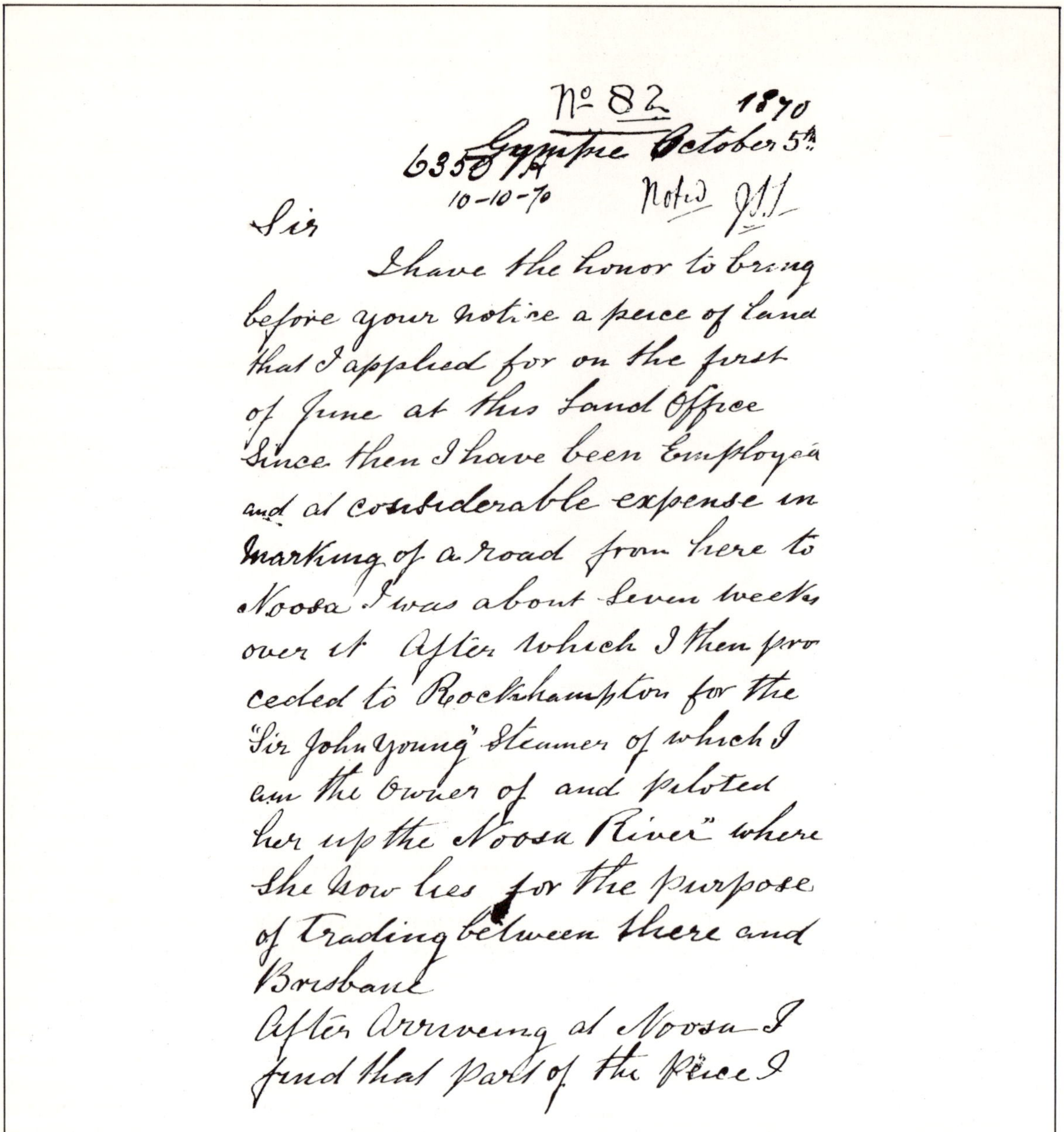

No 82 1870
6350 H Gympie October 5th
10-10-70 Noted [illegible]

Sir
I have the honor to bring
before your notice a peice of land
that I applied for on the first
of June at this Land Office
Since then I have been Employed
and at considerable expense in
marking of a road from here to
Noosa I was about seven weeks
over it After which I then pro
ceded to Rockhampton for the
"Sir John Young" Steamer of which I
am the Owner of and piloted
her up the Noosa River" where
she now lies for the purpose
of trading between there and
Brisbane
After Arriveing at Noosa I
find that part of the Peice I

Walter Hay

First page of a letter from Walter Hay to the Surveyor-General, October 1870. (By courtesy State Archives, Brisbane.)

Soon after he moved to Noosaville he lost his first wife, the former Mary Anne Eaton of Teebah Station, near Brooweena. As his first family grew up, he built cottages for them on his various parcels of land so that they could fulfil the residence clause for gaining freehold. He re-married in 1877.

By now the name of Hay was dotted over the early parish maps of Laguna, Weyba and Tewan-tin. The other landholders south of the river were Goodchap, Hawley and Matters. George Paige Matters was a coachman on the Noosa–Gympie run when he applied for freehold of the 40 acre property he was leasing south of Weyba Creek. It was now 1878, and he had built a three by three and a half metre weatherboard cottage, used as 'a seaside residence' by his family for two months of the year, while the applicant 'resides there three nights a week all the year'.

The 91 acres between his place and the Weyba Creek entrance was also a holiday property, owned by Joseph Aquila Hawley, merchant, of Gympie. These were the first 'weekenders' in Noosa Heads.

In 1882, Walter Hay senior and his son, also Walter Hay, applied for 640 acres 'twenty chains north of Coolum Humpy, on the sea shore', to be transferred from the original selector, Grainger Ward. This was Coolum Hill (not Mount Coolum), now the site of valuable Coolum Heights development.

Walter Hay was now actively speculating in land, holding about fifteen separate parcels at dif-ferent times, acquiring some from the original selectors and selling others to newcomers. His name was well known in the Gympie Lands Of-fice. Perhaps this is why a James Forsyth appears as the nominal owner of Lot 171, the present site of Hack's banana farm, which formerly stretched from the quarry reserve on Laguna Lookout hill right down to upper Hastings Street, and included Morwong Drive. The adjoining land was covered in large cypress and hoop pine, which Walter Hay proceeded to cut for timber.

When Hay applied for this lease, No. 1644, with a frontage on the Noosa inlet behind Hay's Island and stretching for 61 acres up the hill past the pres-ent Noosa Heads Hotel, he was residing on Lot 171 with Forsyth. It was as Forsyth's agent that he had first applied for the land in 1876, 'at the N.E. corner of Section I of the Village of Noosa'. It was 1880, and there was a dwelling house of four rooms of hardwood and pine, with veranda in front. Four acres of trees had been ringbarked and eight acres were under cultivation for corn,

potatoes and other crops. Five cattle and four horses were owned by Hay.

As they had not complied fully with the residence clause, the deed of grant was not issued. Hay applied again in 1882, by which time the homestead was valued at £220. Forsyth, giving his address as 'Laguna Bay', had asked for a piece of land adjoining in the Aboriginal reserve, as his own land was too stony for him to put up a fence! He pointed out that the mission to the blacks had 'fallen through' and the land 'has not been used for years for mission purposes'.

In fact there was another reserve set aside on the east side of Weyba Creek for the Reverend Fuller's mission. This too was taken back. It is possible to see from the Lands Titles records how, in a few years, every bit of waterfront, every sacred site, every source of fishing and of fresh water was in-exorably taken up by the white men. This was done with no regard for the black men who thought they had 'owned' it for thousands of years; and entirely without compensation — ex-

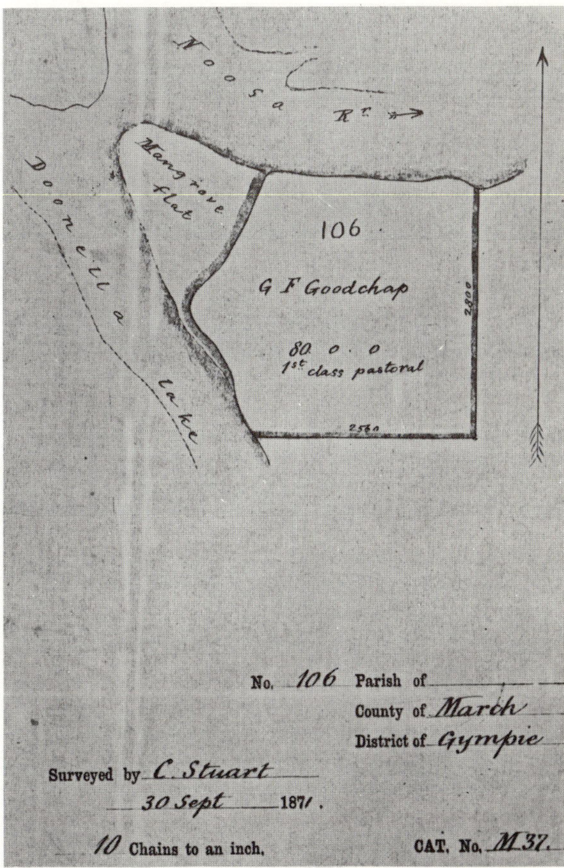

Plan of Goodchap's estate, the first freehold south of the river. (By courtesy State Archives, Brisbane.)

Parish Map of 1900, showing Noosa Town Reserve, parishes of Weyba and Laguna. (By courtesy State Archives, Brisbane.)

cept for the ludicrous brass plates that proclaimed the poor bewildered old men of the vanished tribes to be 'kings' — kings without either land or subjects.

The piratical process is still going on, though now it is the white residents who are losing their land rights. What has become of the public reserve on the west bank of Lake Weyba, near Ennie Creek? What of the 70 acres of 'public purposes reserve' at Munna Point (gazetted in 1884) and now partly a council caravan park, a small sandy waste, and a large canal development enclosed by a high barbed wire fence? Only the Noosa Woods (R.336) seems to have been kept intact and, so far, 64 acres of paperbark swamp — an overflow area at spring tides — between Weyba Creek and Keyser Island (also a reserve).

Another piece of property in the Hay family adjoined Burgess's on Lake Weyba. Walter Hay's widowed mother had come down from Maryborough. She had married again — a Mr Christopher Atkinson — but was widowed a second time. Isabella Atkinson wrote to the Lands Department in 1886 with a request 'to allow my son Walter Hay to act as my agent at the Gympie Land Court as I am over seventy-eight years of age and I fear the journey on the rough road between Tewantin and Gympie would be too much for me to stand'.*

Lot 1651 was duly granted to her, and Mrs Hartley (who went to live on the lake shores in 1896) remembered her as still living there at a great age.

The other big Hay selection, No. 1653, surrounded the Burgess estate but consisted mainly of low heathland. Its northern frontage, however, was to the surveyed 'Golden Beach road' and included the Noosa Junction shopping area and much other valuable real estate today.

The 1100 acres were forfeited about 1886 when they were bought from the Lands Commission by Charles S. Miles. Walter Hay had asked for a reversal of forfeiture on the grounds that he'd had a bad year, with rain keeping guests away from his Bayview guest house and spoiling his vegetable crops. He asked for time to pay the arrears of rent on the lease. However, the Lands Commissioner refused this, appending a dry note to the effect that 'W. Hay can hardly be classed as a farmer, as he owns a Boarding House and also extensive land

holdings in the area'.†

Walter Hay received the freehold of his Noosa Heads portion in 1888, when he gave his occupation as 'Boarding House Keeper', and listed improvements including 'a 6-roomed cottage with veranda rooms, of hardwood and champherboard with galvanised iron roof' — the first Bayview. He had another cottage higher up the hill, near where the old water tanks can still be seen beside the public footpath, amid a jumble of orange bignonia climbing the trees — a creeper that once draped his front veranda.

Hay's Island, in the shelter of the Noosa sandspit, was so named after the original Walter Hay. It apparently remained Crown Land right up to the time the Lands Commission leased it to the latest developer, Cambridge Credit, who renamed it Noosa Sound. Because it was low-lying and partly inundated at high tide, it was not included in the Noosa town survey, except as a dotted outline marked 'Island'.

Walter Hay cleared some of the higher ground, above high-tide level, and established a market garden there. Fresh water was available as surface water or from shallow wells, seeped down from the high Laguna Hill, and no doubt the soil was rich from silt and bird droppings, as the island had a dense and varied vegetation.

Later, his son Dan, who was living on the island, went with Captain Alexander Goodall of the *Culgoa* to the rescue of a man whose boat had overturned near the North Shore. Both he and Captain Goodall received a medal for bravery. Goodall was said to have saved more than a dozen people from drowning.

There were beacons of some sort on the island to guide ships over the bar as early as 1869. When Walter Hay was appointed official Light Keeper by the Department of Harbours and Marine, the beacons were acetylene or kerosene lamps that could be lined up at night. Up to thirty years ago there was a gas generator on the island for the acetylene. Later, the bar became more dangerously shallow, and the lights of houses confused shipping, so the 'Noosa Harbour' was closed.

Percy Hay, a son of Walter Hay's second wife, Susanna Jane Bull, became the next harbourmaster, followed by George Hay, who kept the daylight marks painted, and repositioned them as the channel changed.

As Light Keeper, Walter Hay was allowed to build a house on the island, and he held this appointment up to his death in 1907 (although by then he left the work to his sons). The Hay family grew enough vegetables on the island, and tropical

<hr>

*Lands Department, Files 1868–96, Queensland State Archives, Brisbane.

†Ibid.

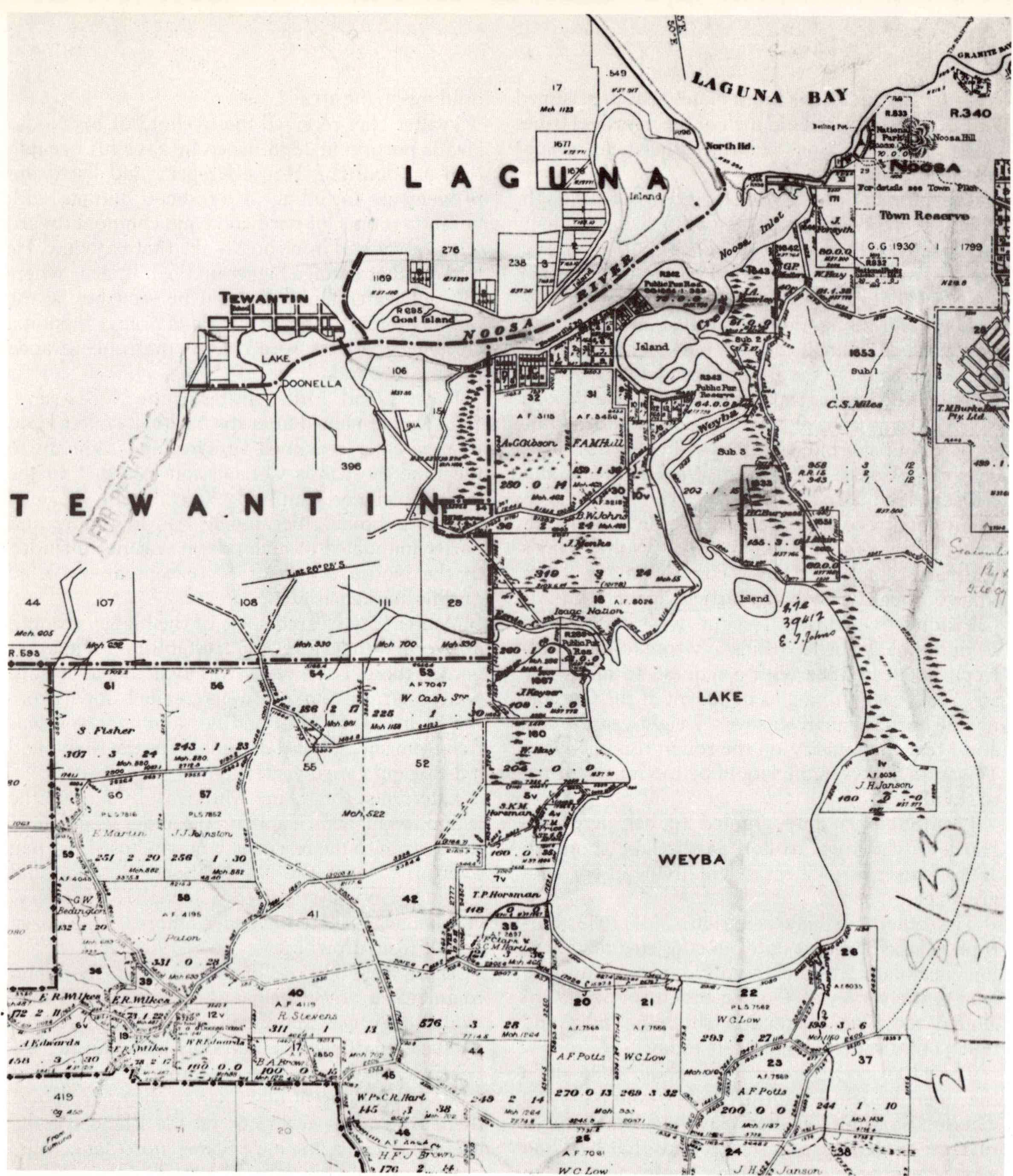

Parish map of Weyba, 1931, showing early Noosa National Park in two sections, and gazetted recreation reserves. The park was gazetted in 1930, the enlarged park of 945 acres (382 hectares) in 1972. This included the Noosa Town Reserve, and land formerly held by the council at Hell's Gates. The freehold owned by T. M. Burke at the other end of Alexandria Bay is now included in the park. (By courtesy Department of Lands.)

fruit on the sunny slopes of Laguna Hill, to supply all of Noosa Heads and its two guest houses.

Of the numerous family of Hay's first wife, the young Walter (Walter Aberdeen Hay) was a good stockman who took up land at Brooweena, where his mother came from, and ran cattle there. Some of his descendants still live in the area.

4 Pioneers

The local residents, determined to provide a school for their children, asked the aborigines to strip bark from the stringybark trees for a little school, which was built near the big fig tree. One of the children attending . . . was the daughter of Joseph Keyser, the first professional fisherman in the Tewantin–Noosa district.

Ailsa R. Dawson, *Cooloola: Early Chronicles of Cypress Land* (Gympie: privately published 1976)

The towns of Tewantin, Noosaville and Noosa (which includes both Sunshine Beach and Noosa Heads) have seen, in their brief lifetimes, every form of transport from the Aborigines' bark canoes to the modern small planes of Noosa Air. On the shores of Lake Weyba, where the local tribes once poled their frail craft, there is now an airstrip. Visitors can be whisked up from Brisbane in less than thirty minutes.

The first white man's transport in the area was the sailing skiff and the ketch. Then came the paddle-steamer (1870–90) followed by the motor launch (about 1900). One of the earliest of these was Colonel E. B. Reid's *Jannet*, while Mr J. B. Atkinson of 'Cloverdell' on the north bank of the Noosa imported the steel-hulled *Woolumba* from America. *Woolumba*, meaning 'place of big water', may have been the Kabi-Kabi name for Tewantin originally. Mr Atkinson gave 'Sergeant' Brown a brass plate marked 'King of Woolumba', while Colonel Reid gave another to King Tommy, 'King of Noosa, Weyba, Cootharaba, and Tumburrawah' (early name for Tinbeerwah).

The three towns were first settled by sea. This was because the country south of Noosa was inhospitable heath intersected by several large rivers. Bullock wagons made the first tracks to Tewantin from the timber country beyond the lakes. Cobb and Co. coaches began a regular run from Brisbane to Gympie in 1868, crossing the rivers in their higher reaches by barge or ford.

The coaches had leather springs and four sturdy wooden wheels with iron felloes shrunk on to them. In their wake came the saddlers, blacksmiths and wheelwrights who would service them. Hotels and staging houses sprang up on the route. Myles' (later Tait's) Royal Mail Hotel at Tewantin put up passengers and refreshed drivers on 'the Short Cut' route to Gympie.

The dispossessed: 'King' Tommy, 'King' Brown (Sergeant Brown) and his lubra Susie in 1880, showing the brass plates donated by Colonel Reid and J. B. Atkinson. (By courtesy Bill Griffiths, Tewantin.)

The first mail run between Tewantin and Gympie started on the rough bush track blazed by Walter Hay. For years the mail was carried by pony express, in saddlebags. Mr John Hall (who had come from the U.S.A. to the Gympie goldfields, and then brought his wife and two-year-old son to Tewantin) was the mail contractor. His daughter was the first white girl born in the town, his wife the first woman to be buried in the cemetery. Mrs George Clifford, formerly of the House of Bottles (which houses a large display of old and rare bottles at Tewantin), was his grand-daughter. Her mother, Mrs William Hall, used to play with the Aboriginal children under the old fig tree.

Mrs Clifford recalled hearing the story of a Tewantin woman who ordered a new hat with ostrich-feather plumes from a shop in Gympie. There was no way the hatbox would fit in the saddlebags, so the resourceful John Hall placed the hat on his head for the journey. When he arrived he caused quite a sensation, but the owner of the 'used' hat was indignant.

A regular mail-coach made the run from 1880,

The Jannet, one of the earliest motor boats on the Noosa River, circa 1902. Aboard are the owners, Major (later Colonel) D. E. Reid and Mrs Reid. In the background are Laguna Hill and the sandspit on the North Shore. (By courtesy Mrs E. Lindley, Buderim.)

This ancient, scratched print shows the motor launch Como on the Noosa River in the early 1900s. (By courtesy Mrs N. Horton, Tewantin.)

when Cobb and Co. applied for permission to erect stables on Crown Land, 'ten miles from Gympie on the Noosa Road'.*

The railway came through slowly, first to Caboolture, then to Yandina, and finally to Cooroy and Gympie. In 1891 the last coach made its run, with Charlie Crank driving.

W. Martin's Halfway House at Cooran was a stopping-place on the route to Gympie. The Martins' son, Dan Martin, the father of Mrs Frank Bickle of Tewantin, later built the Tewantin Hotel overlooking the river by Lake Doonella. This hotel, with its coffee lounge, fish suppers, famed hospitality and superb view, was known as 'The Pleasure-Seekers' Home', with full board at only £1 per week. Like so many other wooden hostelries, it was burnt down in the 1940s.

---

*Lands Department, Files 1868–85, Queensland State Archives, Brisbane.

Tait's Royal Mail Hotel, later bought by John Donovan of Laguna House (who put his wife's relatives, the Gibneys, in to manage Laguna) was equally famous. It, too, was burnt down, but replaced by the present building with its old mail-coach in front.

Tewantin was still sixteen kilometres away from the nearest railway station at Cooroy. About 1912, the hotelkeepers of Tewantin combined with the Eumundi Progress Association to press for a branch line to be put through from Eumundi. A survey was carried out and promises were made by members of parliament; but when World War I broke out in 1914 there was no money for local railways, and the line was never built.

It says much for the attractions of the area that a place cut off from the centres of business, and kilometres from the nearest railway, became so popular. Yet so many managers from the Gympie mines built their holiday houses at Noosaville, on

The last Cobb and Co. coach (centre vehicle) to make the run from Brisbane to Gympie. The driver is Charlie Crank. This picture was taken the day before the opening of the railway to Gympie (extension of the North Coast Railway, 1891).

Dan Martin's Tewantin Hotel, 1912. Charlie Crank sits on the rail (right) behind his horse, which is tethered to the veranda. Neck to knee bathers hang to dry over the upper balcony and there is not a car in sight. (By courtesy Frank Crank.)

Still truly rural: sheep outside the first Royal Mail Hotel, Tewantin, circa 1900.

'Refreshment Rooms' at Tewantin, 1910. (By courtesy Caboolture Historical Society.)

An outing in a rowing boat on Lake Cooroibah before the days of outboard engines. (By courtesy Mrs Gilliland.)

Taken from the Tinbeerwah Lookout: the Noosa estuary from above Tewantin in the 1920s (before Doonella and Weyba Creek bridges were built). The *photograph shows how Noosa Heads (in distance) was isolated, all communication being by boat. The ferry* Emerald *took passengers across to Noosaville.*

the calm and lovely lower reaches of the Noosa, that the road alongside became known as 'Gympie Terrace'.

Others retired there, bringing their wooden houses with them and setting them on stumps on the river bank. There they could sit in the sun on their northward-facing verandas, watch the boats go by, tend their crab-pots, and cast a line for bream from their private jetties. All had boats, for there was no way either to Tewantin or to Noosa Heads except by Jack Tedford's ferry, the *Emerald*. Men like Colonel Reid, manager of the Scottish Mining Company's operations in Gympie, Freddie Kearton, an engineer, and Richard Parkyn, a mine manager, settled there. Soon there were eight houses along the front, and four houses and a shop at Munna Point, where there was freehold land right on the water's edge. Mr Tommy Williams lived there first, with his boat, the steam-propelled *Maggie*. Later, John Parkyn, son of Richard, took over the Point with a boat-hiring business and store. He was the first to clear the paperbark forest there.

There is still some private river-front land near Munna Point, but fortunately the Lands Department of the day would not allow the whole bank to be alienated for private use. A road and foreshore reserve were surveyed in front of the freehold and leasehold blocks for the length of Gympie Terrace, so that this sandy river bank with its shady trees is available to all: a favourite place for safe swimming, picnics, fishing, and launching boats.

Up-river from Parkyn's was William Gibson's, then John Ely's, the Massouds', Metcalfe's and Ferguson's. Colonel Reid's house 'Kilfinan' stood where the Noosa River shopping centre is today. Ned Ely and Alex Gibson, of the next generation, each had a fishing boat. Mrs Ned Ely, who was Annie Durham and who lived at Noosaville, opened a pie and tart shop on Gympie Terrace, for she was 'a light hand with the pastry'. During World War I she was cook at Laguna House. She recalled when, with over two and a half metres of water over the bar, both sharks and porpoises were common in the river.

Alex Gibson's method of getting his Ford T buckboard across the river to the north shore: straddling two boats. The boats were towed by a launch. (By courtesy Mrs N. Horton.)

Lola and Howard Parkyn pose outside their grandfather Dick Parkyn's house at Noosaville with the jaws of a three metre shark, circa 1915. (By courtesy Howard Parkyn.)

Giant groper caught at Munna Point by Alex Gibson (left) and Howard Parkyn. At 206 kilograms, it was then a record for Queensland. (By courtesy Howard Parkyn.)

'Sharks? Heavens above!' cried Mrs Ely, then eighty-five. 'The sharks used to go past like troops of soldiers . . . . Outside one of the shops they had the jaws of a twelve foot shark when we were children. My brother Tony Durham crawled through them.'

Howard Parkyn, who, with his brother Cloudsley (Jack), took over his father's business and ran the *Miss Tewantins* I and II as tourist launches for forty years, can remember posing outside his grandfather's place with the jaws of a three metre shark caught by John Parkyn at Munna Point.

They were said to go right up to the beginnings of fresh water beyond Cootharaba, to get rid of barnacles. One of the Massoud brothers, when putting out a net, was chased by a two and a half metre shark in Lake Como. He said his brothers

were laughing so much they could hardly pull him into the boat.

There were giant groper in the river too — all cleared away now by spear-fishermen — which occasionally took a dog, but mostly lived on crabs. One, caught by Alex Gibson and Howard Parkyn, weighed 440 lb, then a record for Queensland. It had made a recent meal of crabs, which started to walk away, but they had turned bright red as though cooked by the fish's digestive juices.

The little war memorial at Tewantin, under the big fig tree, bears the names of many of those early families whose sons went to fight in two world wars, some never to return.

George Burgess, back from the first war, and aged by his experiences though he was only twenty-one, stayed for a while at Laguna House. He was suffering from shell-shock. Later he went fishing with Ned Ely, using nets of Irish linen (no nylon in those days) two and a half kilometres long. They would catch hardygut mullet that come into the river to spawn, bringing in 140 to 150 cases with each haul.

The late Edgar Johns, of Johns' Landing on the Noosa River, was another who talked to me of the old days. He was in the Australian Light Horse during World War I, and returned with the Distinguished Conduct Medal.

He remembered when his father had the Restdown Orchard at the Landing; he employed Pacific Islanders — then called Kanakas — to work in the orchard at planting and chipping weeds round the fruit trees. They had been imported to labour in the Queensland canefields where the work was supposed to be too much for white men in the summer. But Edgar related:

As for cutting cane, it wasn't hard work if you were used to it. I've cut cane myself — I've cut scrub, split fence and rails, and done most bush work. The worst thing about the bush is not snakes, nor the heat, but the gympie-gympie. That's the stinging tree they named the Gympie scrub after — that's what the blacks called it.

It has a big leaf with fine, hairy thorns, that fine you can't hardly see them. I was stung all over my hands up at Cootharaba. They're frightful things. Every time you wash your hands away she goes again, so that you feel you never want to have another wash! Vinegar essence eased her down a bit; but it took weeks to wear off.

Then of course there were leeches, and mosquitos, and taipans . . . .

By the time he returned from the war, most of the Islanders had been forcibly repatriated, and white men found that they could work in the tropical sun after all, even as far north as Mossman and Mackay.

Sugar was grown for a while near Cooroy Mountain, but frost killed the crops. There was a juice mill there — the Wilkins' — at which another pioneer, Bill Ross's father, Ben, worked for a while. Bill's half-brother Clarrie Ross was the first male child born in Tewantin. The late Bill Ross remembered:

Many a time I've walked out as a kid to get mangoes from the Wilkins' old place. The ruins are still lying about there. My father dug a hole in granite for water — barred it out of the rock, just with a crowbar. It's still running. It's crystal clear and perpetual . . . from this side of the mountain.

Bill and Clarrie Ross worked for a time for Charlie Crank on his cattle property. (Charlie had retired from driving coaches.) Later they had a butchering business and slaughter yard in Tewantin.

Charlie Crank, who later was chairman of the Noosa Shire Council for sixteen years, came from Nanango with his father, Richard, when he was eleven years old. They settled at Tewantin in 1876. He worked later as a blacksmith and wheelwright. He also worked as a stockman for the first Walter Hay, whose extensive stockyards stood on the river bank at Hilton Terrace, Noosaville. Later he took up a lease to the west of Lake Doonella, from Cooroy road right down to the lake — where Crank's Creek runs today.

His son Frank, now in his eighties, used to run cattle both north and south of the Noosa River on grazing leases. Then, when ticks became a problem, he gave up cattle and worked a team of horses. He got work road-building for the country shires, with a team of six horses with scoops and rollers.

Frank was also a timber contractor, and used to ford Lake Doonella and Weyba Creek to get to Noosa Heads and pick up pine logs from Walter Hay. When the two wooden bridges were built in 1929 he put in a tender — it was all done by word of mouth in those days — but his father left the council chamber and would not vote. Another man got the job but went broke. In the end they asked Frank Crank to take over, supplying the timber and contracting for the bridges. 'My horses were good,' he said, 'and you had to have horses for a job like that.'

He said that 'Tommy Dodd' and 'Agnew', sons of old King Tommy (whose Aboriginal name was Tuppernywoe) used to work for his father.

Old King Tommy was a great man. And his sons were good workmen, good dependable men. Yes, I've seen a hundred blacks having a mad corroboree, just near our place. Someone in the town would give them rum, and they'd put on a show half the night . . . . The noise was terrible!

The old men's skill in removing bark had been pressed into service for roofing some of the early buildings in 'Goolloi Street' (now Poinciana Avenue), and this was how they got the rum.

Bill Ross ran away from school when he was only eleven, and one of his first jobs was to go out with the Aborigines to get bark to roof the coach sheds for the Royal Mail line of coaches, run by John Myles from Tewantin to Cooran:

We're loading bark on to the spring van, about three mile out. We'd be going along, with a gin following behind, and these two old blackfellers would say to me, 'You got the beer, my son? You got the lum?' and I'd say, 'No, Father,' (I always called the old men *Father*) 'I haven't got the rum.' Because if I'd given it to them first they wouldn't have helped me, and I was too small to load the sheets of bark on my own.

After we'd got the bark loaded they said to me, 'You tell Mr Paton he no send out the lum.' I said, 'Wait, I'll have a look in here.' (The rum was there all the time.) Then the old feller pulled his eye down like this with one finger, and he said, 'You got the bloody lie, not the nice clean eye.' Because I'd told them there was only beer. Beer was all right, but rum would fire the black man up at once.

Bill Ross died in Tewantin in 1975, aged eighty-one. When he was a boy there were still half a dozen Aborigines left in the district besides the two old 'Kings': Tommy with his sons Agnew and Tommy Dodd; Brown with his lubra, Lucy, and their son Freddy Brown, a young fellow of fine physique; Willy Dunne and his lubra, Emma.

The old men showed him how to find 'sugarbag', the honey of the little native bee or *gilla*, by following the flight of a homing bee; and warned him always to watch ahead of his feet, because 'might-be that bigfeller snake sit down there'. Death adders were common, and King Tommy always had his tomahawk ready.

They also showed him their skills in spearing mullet and tracking flathead in the shallow waters of Lake Doonella and Lake Weyba. The flathead were thick, especially by the old Weyba Bridge ruins. They would follow them up, walking against the tide so that any mud or sand would be carried away by the flow. When they had tracked a fish by the mark he made on the bottom, one man would go ahead with a towrow net while the others frightened the fish into the net. They used to sell their catch in the town for a penny a fish. Said Bill:

Then they got very old; but the publicans were good to them — they had a camp down by the old iceworks below the steep bank where the old fig tree is. Mr and Mrs Dan Martin, Mrs Bickle's parents, and Mr and Mrs John Tait, would send us kids down with a billy of soup, or some bread and meat.

When a tribe would come down from Gympie for a feast in the old days, there might be twenty of them camped out there in what was called Henry's paddock. They'd all be done up, and they looked as wild as scrub hens! We didn't understand the words of the songs, but the rhythm of it would get you . . . . It was like a ballet you'd see on the stage. And us kids loved it.

They're all gone now. The local people didn't want them to go, but the police and the government rounded them up, and sent them off to Murgon, or Barambah.

Bill Ross remembered a big hammerhead shark in the river, taking the deadwood of Dan Hay's boat in its jaws and trying to turn it over. He also recalled how he and his brother helped to raft whole logs from the Kin Kin scrubs. They were hauled to Kin Kin Creek by big draught horses and floated down to the lake. In the lake they would be chained together in rafts of up to 220 logs. If laid end to end, they would have stretched for nearly a kilometre.

It would take about a fortnight to bring them down to Tewantin. You'd take a rope out with an anchor on, about one hundred yards, and a winch on the raft. So you'd winch up to that, then take another anchor out, and winch up to *that*. It was called 'cadging'. Then, if rough weather got up and the raft broke apart, you'd have to gather them together with a rowboat.

In the days of the McGhie/Luya mill, sawn timber would be brought down in barges towed by three small 'droghers' (paddle-steamers). After milling operations were moved down to Dath/Henderson's at Coolloy on the lower river, whole logs had to be moved: hoop pine, kauri, and cedar. The boys worked for five shillings a day. Later, the rafts were towed by the motor-boat *Dawn* in much less time. The skipper of *Dawn* was the late Alex Gibson, whose father, William Gibson, took up land in Noosaville, between Hilton Park and Keyser Island.

The old bridge over Weyba Creek, circa 1900. The bridge burned down in the early 1900s 'owing to bushfire'. Some thought it burned because it was too low to allow fishing boats underneath to the Weyba Lake fishing grounds. Oysters as well as fish were exported from Tewantin.

Jack Tedford senior was a fisherman who invested in Gympie gold shares, and was able to open a butcher's shop in Tewantin in the 1880s. At about this time, Richard Crank and William Clark certified in the Gympie Lands Office that Jack's brother, George Tedford, had completed his residence on a neighbouring lease of 300 acres. Here the Tedfords ran beef cattle, but, with rump steak at twopence halfpenny a pound and long distances over poor roads to deliver the meat, the venture failed. Jack returned to fishing with his young sons, Jack junior and Bill.

Jack and his brother saved enough to buy a boat and an engine, and for many years Jack Tedford ran the *Emerald* as a ferry between Noosaville and Tewantin. He used to dredge oysters from the river-bed for the Moreton Bay Oyster Company. When his boat was attacked by marine worms, he filled the bottom with cement and sailed her down to Brisbane for repairs.

As well as individual settlers, an agricultural settlement on communal lines was begun at Lake Weyba in 1894.* This was called 'the Wooloongabba Exemplars' and comprised people displaced by the great Brisbane floods of 1893. They were city folk and almost starved in the midst of plenty; though they had an area of lake frontage, they seemed unable even to catch fish or crabs or find oysters. The commune was soon abandoned, for it had no permanent supply of fresh water, the lake being salt. Some of the settlers later did well at Skyring Creek, in a group called the 'Protestant Unity'.

The Unity group was another experiment in collective farming. The government set aside 3000 acres to be farmed jointly by twenty-six members and their families, the government finding food and tools at first. The group continued for two years but then broke up. The land was divided among the members, and each farmed his own piece individually.

Then, in 1904, the 'Richmond Settlers', a group from the Northern Rivers area of New South Wales, began felling the big Kin Kin scrubs, turning the rich soil into dairy pastures. A road was constructed across the hills and the town of Pomona was established. This was later to become the headquarters of the Noosa Shire Council, which was separated from the Widgee Division in 1910. The district's first newspaper, the *Noosa Advocate*, was published in Pomona from 1909.

Among the early identities of the area were the

The late Annie Chesworth Krafft, nee Burgess, the first white child born on Lake Weyba 100 years ago. (By courtesy George Burgess, Noosaville.)

elder Massouds. They arrived from Lebanon, speaking little English but determined to make good in the new country. They travelled on foot about the district hawking clothing, materials and other goods at farmhouses. Then they set up a shop and soda-fountain next to Mr Ely's wooden house (lately pulled down) on the river bank at Noosaville, and sent their children to school in Tewantin.

The mail was delivered by river three times a week. Jack Tedford was the first to run this service and later it was taken over by Percy Hay in the *Ara*. Supplies came from Brisbane by sea twice a week. Not only mail and stores travelled by boat: most of the children, too, used water transport to get to school. Some, like Frank Crank and his brothers from the Cooroy road, could ride horses to school because they were on the right side of

---

*Shown on the parish maps of 1896 (Parish of Weyba).

42

Doonella Lake and its tributary creeks. Others, like George and Annie Burgess, would have had a very long row. 'I hardly went to school,' says George Burgess, another old resident of Gympie Terrace. 'It was too far from where we were, right round on the far side of Lake Weyba . . . . My sister Annie was the first white baby born in those parts, in 1882.'

His father, George Chesworth Burgess, took up land on the north-east corner of Lake Weyba, not far from its junction with Weyba Creek. He built his house on high land overlooking the lake and Jubilee Island. It was a big place built of pit-sawn timber, with a bark roof covered with galvanized iron. Its site is now Weyba Ranch, where American quarter-horses are bred.

Burgess senior was a remittance man, with a regular income from wealthy relatives in England. His son, now nearly ninety, remembers his father, as always immaculate in white duck suit and planter's hat, walking about overseeing the twenty-odd Kanakas he had clearing and chipping weeds in his orchard:

You never saw him soiling his hands . . . . He used to keep pigs, and would let them kill one for a feast. He planted mulberries, and Isabella grapes, in fact fifteen varieties altogether; but in a very wet season they all got the mildew.

It was a great place for keeping bees. I had forty hives of paperbark honey, and wildflower honey in the spring. There were wild brumbies in the scrub, wild ducks on the lake, and dingoes were always killing the fowls. The plain turkeys were thick . . . . wood pigeons and flock pigeons passed overhead in thousands. We milked goats, had home-made butter and clotted cream with honey, bread baked in a camp oven, and mulberry wine in five-gallon kegs . . . . And of course there were always fish in the lake.

Percy Hay also kept hives for honeybees in the paperbark swamp below Bayview (now drained and built on) and guests would always take home a pot of the delicious paperbark honey. As unofficial harbourmaster and official Light Keeper, he tended the beacons in his boat the *Ara*.

The elder Walter Hay's grand-daughters, Polly (Mrs Greber) and Rose (Mrs Cobby), lived in the Noosa area for many years and died only recently. Polly's daughter Olive became Mrs Charlie Freeman, and Olive's son, Kevin Freeman, still lives on part of the land taken up at Noosa Heads by his great-great-grandfather in 1875.

Olive Freeman, Walter Hay's great-grand-daughter ('full of life — a bit of a tomboy', as her contemporaries remember her when she was a girl) used to row a heavy dinghy on most days, all the way from the Heads to Tewantin. She would stop to pick up the Massoud children on the way. Sometimes, when the tides were against them, they would not arrive till ten or after, and would

The Massouds on the North Shore — then an uninhabited world, except for sea birds and fish. (By courtesy Mrs N. Horton.)

sneak in at recess time with the others. In summer it was often raining heavily. They would arrive soaked from their journey in the open boat.

As the Massoud boys grew older, Bill took over the oars and would row on from their place. His sister, Maisie Monsour, remembers how, when they were younger, the boys insisted on landing on Goat Island to pick wild cherries. When they refused to 'come on', Olive rowed off and left them marooned, but she came back and got them later.

John Donovan bought the *Shamrock* motor-launch to collect stores and passengers from Tewantin. Soon, young Lionel Donovan, aged four when his parents took over Laguna House, was speeding past the others on the way to school.

Mr Massoud senior then bought the *Riverlight*, and got the contract for delivering schoolchildren from all along the river, with young Bill as the skipper. The combined school- and mail-run was taken over later by Mr Parkyn. Bill had the contract for running mail and stores to the lighthouse at Double Island Point after he grew up, and often took passengers as well. He had a series of old

Dodge tourers. If the radiator boiled he would fill it with seawater, and if the engine got wet and wouldn't start he would pour petrol over it and set it alight. His cars did not last long.

It was Lionel Donovan, the first child in Noosa Heads to go to school by motor-boat, who was to lead the development of motor transport in the area. He was 'mad about cars' when he grew up, and still had his original number plate, No. 141, and his first driver's licence.

His fleet of modern tourers, sedans and charabancs used to meet the trains at Cooroy on a regular schedule in the thirties, and transport passengers to the hotels in Tewantin, and to Noosa Heads guest houses.

An advertisement in the *Noosa Advocate*, about 1925, for the Royal Mail Hotel mentions 'Good stabling and horse paddock, plus spacious motor garage. Best of forage.' It was the transition time between horses and the horseless carriage. The Fig Tree Store and Cafe advertised 'Soda Fountain Drinks and Ices.'

In 1931, the year the big flood in the river nearly came up to the front veranda, Mrs M. Dwyer was

The late Lionel Donovan's fleet of cars and charabancs, 1938. These were used to meet passengers at Cooroy Railway Station and convey them to and from Tewantin and Noosa. Lionel Donovan is on the right. The cars, modern then, are a Buick 1929 (the oldest), three 1936 Chevrolets, a 1932 Chevrolet and a 1934 Ford (second from left). The fleet is parked outside his Mayfair Picture Theatre, which was later burned down. (By courtesy Mrs Beth Smart, Maleny.)

San Elanda guest house at the end of Pelican Street,
Tewantin, overlooking the river. (By courtesy Frank
Bickle.)

Tewantin, before the new post office was built, showing
San Elanda guest house, left, the old fig tree, the new
Royal Mail Hotel and, right, Mt Cooroy. (By courtesy
W. H. Robinson.)

Ruins of the Mayfair Theatre, burnt down in 1940 and rebuilt on the same site. Now the site of the shopping ar- *cade in Poinciana Avenue. (By courtesy Olive Freeman.)*

running Elanda House (later San Elanda) at the corner of Pelican Street, overlooking the river, with 'every home comfort'. Mrs John Donovan took over the running of San Elanda guest house after she was widowed.

The old town of Tewantin, left behind in the mad explosion of motel and restaurant development for the tourist trade at Noosa Heads and Noosaville, has retained much of its charm. There are still big trees and grassy slopes down to the wide bend of the river (though the old fig tree has lost half its branches). The grand-daughter of both J. B. Atkinson and William Gibson, Mrs Norma Horton, still occupies the old wooden house on the river bank where she was born, next door to the imposing new council chambers.

Though its construction was opposed by many ratepayers, this expensive building on a prime riverside site, comprising the old council caravan park and some private land, has become a social centre of some importance to the community. Art shows, including some prestigious loan exhibitions, are held regularly by the Noosa Gallery Society in the spacious art gallery, and the shire library occupies the ground floor overlooking the river.

Though a new office and shopping complex is being built in Poinciana Avenue, the old town still contains some weatherboard houses on stumps, the typically 'high-set' Queensland style, interspersed with new bank buildings and the shopping arcade where the old Mayfair Theatre stood before it burnt down. (The ancient Theatre Majestic at Pomona has taken over for those who want a change from bland drive-in cinema fare.)

The Tewantin Post Office won an architectural award in 1961 for its functional design, but is now rather dwarfed by the council chambers set in their landscaped grounds. The hotels and guest houses that used to stand on almost every corner have shrunk to one, the Royal Mail.

And gradually the old identities, pioneers and sons of pioneers, are going too: the late Bill and Clarrie Ross, Jack Tedford, Jack Parkyn, John and Lionel Donovan, Charlie Crank, Ivan and Bill Massoud, Arthur Finney and Alex Gibson. . . . Even the pelicans, which gave Pelican Street its name, have left for Noosaville, where they can beg fish scraps around the *Laguna Belle* and the tourist jetties serving the *Allikat* and the *Cooloola Queen*.

5 The village of Noosa

Noosa Shire has three basic attributes which . . . attract people to the area. These are its beaches, its river, and its national park. These can be developed intelligently, or they can be ruined. Which is it to be?

Dr A. G. Harrold, Secretary, Noosa Parks Association. From an article in the *Noosa News*, 17 February 1977

Fifty years ago Noosa Heads was still a village, with only a bush track leading from the sheltered beach back up the hill, and down the spit to the boat landing at Noosa Woods. It was 1929 before the first road communication with the outside world was completed.

A bridge was built over Weyba Creek in about 1900 but, by 1910, it had been burnt down and was not replaced, so the headland and the beaches both south and north of it remained isolated. Weyba Lake teemed with fish in those days, and it was said that the convenient 'bushfire' which demolished the bridge was started by disgruntled fishermen who could not get their boats up to the lake.

Noosaville, on the river bank between Weyba Creek and Lake Doonella, developed first after Tewantin.

By the turn of the century the population of Tewantin had grown to 148,* while at Noosa Heads there were still only five houses by 1905. Two of these were guest houses: Hay's Bayview and Bainbridge's Laguna House. Up on what is now Motel Hill, the site of expensive modern development, Matters' cottage overlooked the sea. Nearer to Weyba Creek was Hawleys'.

In what was to become Hastings Street, the Ramseys' wooden house stood next to Laguna House, where Pinetrees is now. It was used for the overflow when the four-bedroomed Laguna was full.

In 1906, the progressive John Donovan bought Laguna House and began putting Noosa on the map. He soon had so many guests coming back to sample the marvellous meals of fresh fish, prawns, oysters and crab, and the safe swimming on the beach just opposite, that he brought some wooden

---

*Pugh's Almanac, 1889.

The first graded road down to Noosa Heads beach, through original rainforest (now Motel Hill).   *Photograph by Murray Views of Gympie. (By courtesy Mrs Gilliland.)*

Old Bayview, owned by Walter Hay, circa 1900. (By courtesy Olive Freeman.)

Guests at Bayview, circa 1919. (By courtesy Olive Freeman.)

houses down from Gympie for extensions. When the extensions were completed, Laguna House stretched forty-eight metres, with wooden balconies and steps over which bright towels and neck-to-knee bathing costumes were flung to dry.

Appeals to the P.M.G. for a telephone to 'the Village of Noosa' had fallen on deaf ears for years, so John Donovan put through his own line, an 'earth circuit' needing only a single wire, and using natural trees as posts. It crossed Weyba Creek high enough for boats to pass underneath, but low-flying pelicans continued to sabotage the line.

Between Weyba Creek and Doonella Lake, the south bank of the river was all ti-tree (paperbark) forest with an old bush road, and grassy sward above the sandy shore. The river is as salt as the sea to well above Tewantin, except after heavy rain, and all the lakes are brackish; but the underground water-table is near the surface. Residents each had their own shallow well for sweet drinking water.

At Noosa, there was fresh surface water from a small but crystal-clear creek. This was pumped up the hill into a tank and circulated to residents by gravity feed.

The late William Hay, of 'Hay's Cottage', used to run the motor to keep the pumps working and the tanks full; there was then never any water shortage. Today we have severe water rationing, simply because of poor planning by the council.

At Noosa Heads the great sport is and has always been swimming and surfing (fishing comes next). There is less room for sunbathing these

The original Laguna House, Hastings Street, circa 1906. This was a four-bedroomed, wooden building, set in the bush, with wood stove and kero lamps. (By courtesy Lionel Donovan.)

Matters' Cottage on Motel Hill. This was the first house in Noosa Heads (apart from that of pioneer Walter Hay) and is now the site of a multi-million dollar development. (By courtesy Olive Freeman.)

Laguna House, circa 1950, with its fifty metres of road frontage. Parked outside is the Rolls belonging to John Paterson, the 'Vita-sun' man. This is now the site of Laguna Arcade. (Photo by Kevin Freeman.)

Little Cove, circa 1928. The ladies are still in neck-to-knee costumes and so are some of the men . . . but one young lad has daringly bared his chest.

The water's cold! Bathers in neck-to-knee bathing costumes on Noosa Beach in the 1920s.

Laguna Hill was cleared in 1929 for the planting of bananas and citrus. On the left, Freeman's farm Noosa Vale; right, Hillcrest guest house on the site of Bayview. (By courtesy Olive Freeman.)

In the 1940s, before the days of shark netting, big turtles were occasionally seen on Noosa Beach. This one was rescued from above high-tide mark, where it had become exhausted, and returned to the sea. It took three men to carry it. (By courtesy Olive Freeman.)

days, and the surf beach is not as safe as it used to be. Some dangerous rips have developed towards the mouth, so that the lifesavers, keeping watch from the club balcony at the other end of the beach, often have to go to the rescue.

Perfect surf comes when a big swell, arriving from far out in the Pacific, coincides with off-shore westerlies to smooth the chop and hone each wave to a sharp, clean crest. Then each succeeding wave seems carved out of blue and green crystal, too solid and perfect ever to break.

The word goes round like magic. Then the camper-vans arrive in droves, and young men with long, draggled, faded shorts run down to the surf with waxed boards under their arms. They are out from first light until the water begins to darken in the twilight. There has not been a shark fatality at Noosa Heads since 1961, when a surfer

The first lifesaving reel and belt, with a team from the Noosa Surf Lifesaving Club, circa 1930.

The 'lady lifesavers' pose with the first surf ski at Noosa Heads, circa 1936. (By courtesy Olive Freeman.)

The Noosa team during a surf carnival at Noosa Heads beach in the 1950s. (By courtesy Noosa Heads Surf Lifesaving Club.)

The modern clubhouse, completed 1974. With all amenities, including a liquor licence, it occupies a prime position on the seafront overlooking the beach and the river bar.

was fatally mauled as he waded ashore early one morning. Shark netting has made the beach safer, but also entraps many innocent creatures such as dolphins and turtles, which drown in the mesh.

The Noosa Surf Life Saving Club began in a modest way in 1927, more than fifty years ago.

Beginning with only a tent on the beach, by the next year the members erected the first small wooden clubhouse, nearly a hundred metres in front of where the present imposing structure stands. The earlier building was moved back twice, as the sea encroached on the beach.

Photographs of the seafront taken in the 1947–48 cyclone season show the lifesavers' tower removed and lying on its side, and the shelter and dressing sheds sliding into the surf. The present building could not be moved in an emergency: it is a two-storeyed, cement block construction which has become almost a private hotel, with liquor licence, floor shows, and meals, but no accommodation. The building occupies a magnificent site beside the public car park. The dressing sheds are now tucked safely behind what is left of the frontal dune — its top sheared off to make the

Lifesavers on duty on the 'new' beach at Noosa Heads, 1978. It lasted exactly fourteen months. (By courtesy Noosa News.)

Circa 1945. Left, Mrs Chamberlain's and a private house that is now a vacant block. Behind can be seen Hillcrest guest house, now belonging to the Church of England and renamed Halse Lodge. Freeman's farm and banana plantation (now Hack's) is high on the left. Two of the three palms are still standing.

*A quiet village street: Hastings Street, thirty years ago.
(Photo by Kevin Freeman.)*

parking area. Further along, the dune is built on right down to the high-tide mark.

At Noosa Heads today, only the surf is unchanged, and the three tall blue gums where generations of whistling kites have nested below Halse Lodge (formerly Hillcrest guest house). From the lookout on Laguna Hill, as more and more houses and shops, bowling greens and car yards replace the natural bush, you see that Tewantin, Noosaville and Noosa Heads are growing into a monstrous conurbation, relieved only by the river, the tiny Pinnaroo Park, and the green dot of Keyser Island. Hotels and motels line the steep hill, concrete replacing the dense bush, so that every summer the run-off and erosion become greater.

The land south of the Noosa headlands, known as Golden Beach and rarely visited in the early days, had been taken up first by W. Pilcher, J. Woodrow and others — Pilcher's Gap at Sunshine Beach is still on the map. In 1925 the Noosa Shire Council acquired 447 acres in this area at a compulsory auction for non-payment of rates. The council had bought some other freehold land, making nearly 500 acres in all. In 1928, Thomas Marcus Burke, the land developer, negotiated a deal with the council; he was to receive this land in exchange for £11 000, the cost of constructing a road with two wooden bridges from Tewantin to the new estate. It was to be known as the 'Noosa Beach Estate'.

Noosa Heads beach used to be known as 'Laguna Bay', though the township at the spit was 'the village of Noosa Heads'. T. M. Burke was soon advertising his land at 'the new town of Noosa' from the company's office, next door to the Green Gables Cafe in what is now Sunshine Beach. Locals still referred to it obstinately as Golden Beach, or even Coolum Beach, as the only access was via Coolum sixteen kilometres south.

The whole thing was resurveyed and became the first planned, contoured, and zoned beach

The road blazed through the bush from the Junction to
Golden Beach, 1929. (By courtesy Olive Freeman.)

'Opening Day' for the new road built by T. M. Burke
Pty Ltd, August 1929: festive scenes in the streets of
Tewantin. The old building on the right has recently
burnt down. (By courtesy Frank Bickle.)

The old humpy bridge at Doonella Lake, Tewantin, when it was new (1930). In the foreground is one of the boats used for dredging oysters from the lake bed. The photo was taken looking towards the Goodchap Estate, now the site of the Lakes Motel. (By courtesy Howard Parkyn.)

Enthusiastic use of the new road: two cars pass in the dust. (By courtesy Olive Freeman.)

T. M. Burke's two new wooden bridges were raised in a hump in the middle to let boats through. A young pedestrian tests the approaches. (By courtesy Olive Freeman.)

Timber was hand-cut with an axe for the new bridge. (By courtesy Olive Freeman.)

Charlie Freeman's Ford T buckboard, the very first motor vehicle in Noosa Heads, circa 1929. *(By courtesy Olive Freeman.)*

resort on the Sunshine Coast. But the Depression and then World War II intervened, and the estate languished. The very active Beach Progress Association at the Heads took a hand, forcing the council to continue the T. M. Burke road from its junction with 'Golden Beach Road', down to Laguna Bay. They organized a working bee to widen the track, and soon they had a graded and metalled road (no more bridges were necessary).

Noosa Heads was no longer an island. It would never be the same again.

The two wooden bridges were completed in 1929. On opening day, about seven hundred cars of twenties vintage gathered at Tewantin. It was a typically warm and sunny August day. The town was *en fête*, the bridge was decorated, everyone was out in the streets. At 12.30 p.m. the barriers were to be removed. The Home Secretary for the state (Mr J. C. Petersen) and T. M. Burke executives were to drive in the first car to declare the road formally open.

The late Lionel Donovan had a new '29 Chevrolet tourer with a folding windscreen. A friend arranged to sneak over and lift the chain and, with hood down and windscreen folded flat, Lionel drove underneath. He crossed the bridge just before the official opening in front of about a thousand spectators. The escapade was regarded as a joke and no action was taken against him.

The old, humpy wooden bridges have now been replaced after fifty years of use. The old Weyba Bridge, though no longer used unless for fishing from, can still be seen slightly upstream from the modern structure.

At Sunshine Beach, or 'Noosa Beach Estate' as it was called by the developers, marquees had been erected on top of the high sand-cliffs to cater for the official crowd. The Donovans of the Royal Mail at Tewantin were the caterers.

It was estimated that there were nearly four thousand people at the beach that day. The menu was all local: Weyba crab salad, Noosa whiting, Gympie asparagus, Doonella duck, Pomona pork, Widgee cold turkey and Coolum ox tongue being among the cold dishes. The official hosts were the Noosa Shire Council, their chairman, Mr Edwards, the shire clerk and the engineer.

The first car had come to Noosa Heads before the road. Charlie Freeman, who cleared some more of Laguna Hill for a farm and orchard ('Noosa Vale'), had two motor vehicles: a Ford T buckboard and a smart '25 Dodge tourer. The

60

One of the first motor cars in the area. This 1925 Dodge tourer was kept at Tewantin by Charlie Freeman. Mrs Olive Freeman is at the wheel.

Dodge was kept at Martin's Hotel garage, Tewantin, for use when 'in town'; the Model T was used as a farm workhorse and for family picnics, and carted the first surveyors over to Golden Beach. It travelled down to Noosa Heads with its wheels resting on two rowing-boats, side by side, towed by a fishing launch.

The magnificent sandy beach stretching south of Lion Rock for sixteen kilometres (from Paradise Caves to Coolum Cove) has come into its own: now divided into Sunshine Beach, Marcus Beach, Peregian and Coolum.

Houses are springing up like mushrooms along both sides of the Noosa Coastal Highway, with ribbon development spreading south from Noosa Junction. Several very large developments, such as Glen Eden, have been allowed on the seaward side. But one hundred metres back from high-water mark has been declared an 'erosion control district' and there will be no more building on the seafront, except where the land is already freehold.

In some earlier subdivisions the blocks are far too small, like the cramped nineteen- and twenty-perch blocks on Laguna Hill. There are some at Coolum Beach that are only fourteen perches —

'Not enough for fourteen budgerigars to perch on!' — yet high-rise buildings are to go up on the front at Coolum.

Drainage pipes from some new estates caused gullying and wash-aways in the dunes at Sunshine Beach and Sunrise Estate during the cyclonic rains that must be expected every year or so. In one disastrous slip at Belmore Terrace, half the road disappeared and a boundary fence was left hanging in mid-air. In these dunes the foreshore reserve retains a complex vegetation of horsetail casuarina with its blue-grey, drooping fronds, coast banksia with silver-backed leaves and golden candles, myall acacia, water gum (tristania), pandanus palm and paperbark in the damper areas, and large cypress where sheltered from the wind. A profusion of pink, yellow, and blue wildflowers twines through and under them: golden half-guineas, evening primroses, scented fan-flowers, pig-face and purple goats' foot convolvulus.

All this was threatened in 1972 with sandmining for eighty metres back from high-tide mark, all the way from Paradise Caves to Peregian. Fortunately there were sufficient residents, who made sufficient noise, to cause the state cabinet to override the decision of the Mines Minister, Mr Camm, and

Above: *Gabrons of stone and wire mesh failed to hold these steps at Sunrise Estate. (By courtesy A. G. Harrold.)*

Left: *Beach outlet for stormwater at Sunrise Estate after Cyclone Daisy, February 1972. (By courtesy A. G. Harrold.)*

the leases were not granted. This is an increasingly popular picnic spot with Brisbane families up for the day, or for long weekends, when the sea is lined with fishermen patiently casting their lines for little return (except for a few dart or bream in winter). On weekdays it is still possible to walk the beach in the early morning and rarely see another human being.

It was T. M. Burke Pty Ltd, once again, who were responsible for opening up this length of coast, for so long regarded as useless except as a firing range for the Army. In 1952 Mr Marcus Burke, having revived Noosa Beach Estates Ltd, acquired for the company several more allotments about to be taken over for non-payment of rates, making a total freehold area of 610 acres. By 1957, 400 blocks had been sold in the area. There was still no bitumen road, and the only access was by way of the road from Tewantin or Noosa Heads.

The thirteen kilometres of beach frontage, mostly high heathland and dunes, between the southern borders of the estate and the nearest township of Coolum, was still a wilderness of wildflowers and wallum.

T. M. Burke Pty Ltd then proposed to the Queensland government that, in return for building a bitumen scenic coast road from Noosa to the borders of the shire (halfway to Coolum) and the construction of necessary bridges, the company should receive another 530 acres of Crown Land. The proposal went through in 1958, in the form of a Special (Development) Lease to the company (S.L.23793). The final cost was about £135 000. The land is now, of course, worth millions but much of it in the first-developed area, Peregian Beach, was sold at low figures by T. M. Burke before the present boom in North Coast land set in. The Lands Department is still reaping the benefit, as twenty-five per cent of the sale price of each allotment is paid in return for freehold.

By April 1960, the road and bridges were completed and officially opened by Mr G. R. Nicklin, Premier of Queensland. Interior roads in the new estates opened up cost another £31 000.

The only access from the Coolum end to Peregian was still just a sandy track through swampy, peat country. There was talk of a coastal highway all the way to Caloundra but it did not eventuate. However, tenders were at last invited by the government for the piece from Peregian to Coolum, and this was also constructed by T. M. Burke Pty Ltd in return for 520 acres (Development Lease I) under the terms of the new Crown Land Development Act.

Ironically, the company had no intention of promoting Noosa Heads, north of the headland, when it built the road to the Noosa Beach Estate at the northern end of Coolum Beach. The company had no land at Noosa Heads, which now began to go ahead by leaps and bounds while 'Noosa Beach' stagnated. More than half the blocks on the original subdivision reverted to the company or to the Noosa Shire Council for non-payment of rates, and the few houses erected were either removed or abandoned. Today Sunshine Beach, as it came to be called, is largely built over, but it has remained a satellite town to the expanding Noosa Heads. Strictly, the town of Noosa, as originally surveyed and named, includes both

Fifty years after the first road was built, parking and traffic chaos at the Junction, Noosa Heads, 1981. Cars turn in all directions at the entrance to both Noosa Heads main beach and Tewantin. (By courtesy Noosa News.*)*

areas. Confusion reigns because 'the village of Noosa Heads' is often referred to just as 'Noosa', which should apply to the whole shire.

It was not until the scenic coast road was put through from Coolum to Noosa Heads that the Sunshine Beach area, and the new Peregian Beach development halfway between, really took off. The coastal highway all the way from Coolum to Marcoola is called the David Low Highway, in memory of the efforts of Mr Low, M.L.A., who acted as intermediary between the Crown Lands Department, the local councils, and the developer.

The new road immediately became popular, both as a more direct route to Noosa Heads than the one via Cooroy and Tewantin, and as a scenic route winding over gentle sandhills with sudden panoramas of blue ocean framed in drooping casuarina and spiky pandanus palm. Nowadays, on the northern end, it is more often framed in squares of yellow brick; but it is still a glorious colour, contrasting with the deep, yellow beach and the clear, pale blue sky of the most unspoiled length of the Sunshine Coast.

Diving gannets explode in white foam out at sea, terns plunge in at an angle, a white-breasted sea-eagle drifts from the dunes, a kingfisher utters his shrill pipe. Freshwater lagoons lie behind the beach, with blue waterlilies on their wider reaches. It is a reminder of what Noosa Heads beach must once have been like. To have mined it, flattening the curving dunes and killing the native trees and wildflowers, as was done just north of the Noosa River until recently, would have been a crime. It is only people, their feet, their dogs and children, their polluting rubbish, which this beach has to fear: the plastic bait bags, waxed milk cartons, broken beer bottles, rust-proof aluminium cans, and the effluent from hundreds of septic tanks seeping through the sandhills from above.

In 1981 a concerned citizens' group was formed to protest about the number of multiple dwellings being approved by the council in what had been largely a residential area. The whole esplanade and the adjoining Kingfisher Drive had become lined with blocks of units; the Fortune Caravan Park became a unit site, and Glen Eden, a complex of forty semi-detached units, was built right on the edge of the coastal reserve.

In an unsewered area, this sudden proliferation of high-density living quarters was seen as a threat to health and aesthetics as, after heavy use in holiday times, there were overflowing holding tanks and a scarcely salubrious smell.

Land prices have escalated, and speculators buying at inflated rates have crammed as many units as possible on to their land. Others, as soon as they have got town planning consent, sell their unit sites at fantastic prices, for a tax-free capital gain. It has reached the stage where a seafront site for eighteen units at Sunshine Beach was advertised at $1 100 000 — the land alone!

Demand for more water has led to a large, unsightly storage tank being built on high land at Peregian, and the dunes are under increasing pressure from human traffic.

Progress, as usual, has brought its own problems.

6 Progress is a dirty word

Nature is one vast interconnected ecosystem, and the only ultimately successful conservation is management, with understanding, of the whole complex system. . . . We need this for economic success, but also for our own peace and pleasure.

Edmund D. Gill, Deputy Director of the National Museum of Victoria, *Rivers of History* (Sydney: A.B.C., 1975)

Noosaville, on the river bank between Tewantin and Munna Point, settled first by retired Gympie merchants and mine managers, becomes less picturesque year by year. The old wooden houses on stilts are being replaced by brick bungalows, two-storeyed and four-square. This part of Australia has just discovered yellow brick, which is made locally at the Cooroy brickworks. Modern motels and units, restaurants and supermarkets take up the highly-rated land, while among them a few old identities cling to their old-fashioned houses.

Fishing-boats, trawlers with their blue and green nylon nets and prawn beams, are becoming scarce. The sight of rose-pink prawns, freshly cooked in the saline river water, cascading into containers for market, is a thing of the past. Pelicans still congregate from habit at the Fish Board jetty, but the fleet of trawlers no longer returns with catches from Laguna Bay and the outside reefs. The bar is too silted and dangerous for big boats.

Noosaville is more a tourist resort now than a fishing village. Many of the prawners have sold their boats and taken to running trips to the coloured sands, or further up the north shore to the *Cherry Venture*, a large ship grounded in a cyclone. Others take parties up to Double Island Point and to the Bubbler, a freshwater spring on the beach, where the fishing is still good enough for the long trip to be worthwhile. The *Cooloola Queen* and the new *Allikat* make daily trips up-river through the lakes; and there is even a paddle-steamer restaurant, the *Laguna Belle*. Houseboats, catamarans, and fast outboards have replaced the old put-putts which used to wander up and down the waterways with family parties bent on fishing.

In the upper reaches of the Noosa, where only small boats can penetrate, the water becomes fresh, clear, and tea-coloured; in the deeper reaches it is almost black with peat. The old boats scarcely disturbed the reflections, where fallen trees and curving boughs made amazing abstract patterns as the real shapes met their perfect shadows. These days there are more fallen trees, as faster outboards send out a wash that undermines the low banks.

The Kinaba Information Centre and bird-watching platform is set at the top end of Lake Cootharaba, at the junction of the Upper Noosa and Kin Kin Creek. This is a control point for boats as it is one of the entrances to the Cooloola National Park.

Kangaroos and emus can still be sighted in the bush on either side, or even swimming the river. Sometimes a water dragon or carpet snake may be seen sunning itself on a bough as you round a bend. Blue kingfishers dart along the reaches, ospreys and whistling kites nest high in the trees on the banks.

At Harry Spring's Hut, a favourite picnic spot, there is a tiny beach of white sand. Though the river is narrow, it may be nine metres deep in the centre. At Lake Como and Kin Kin Creek, known locally as 'the Everglades', there is an exotic, jungly vegetation with purple convolvulus climb-ing the trees, while thick ferns and reeds line the banks, and the tortuous channels enclose small green islands. It is a great haunt of fish-eating and weed-eating birds and remains, so far, unspoilt.

There is still no bridge across the river to the North Shore, and for a long time Golden Beach, south of the headland, was just as difficult to get to. Those who knew the way would travel over the rough Doonan road from Eumundi, up and down several steep hills, to the north-west corner of Lake Weyba. Then they would cross the lake by boat, and take a foot track across the uninhabited heathlands to the shore.

In the same way, Teewah and the coloured sands became known through fishermen crossing Lake Cootharaba from Boreen Point to Teewah Landing, then following a foot-track through the scrub. Another popular beach for picnics was Dunne's Beach on the lake's western shore, with its clean granite outcrops, small sandy beaches and leaning paperbarks. Whiting and bream were plentiful in both lake and river, and in the surf.

In the early days, too, there were mud oysters lying thickly on the bottom of the Noosa River and in the lakes. They formed a staple diet for the Aborigines. Later, they were dredged for the

The fishing was great. Trolling for mackerel with a hand line in Laguna Bay.

Brisbane market and, right up to the 1930s, were sent down in bags to be fattened up in saltwater tanks. Then, a worm infestation made them unacceptable and the trade died; so did the oysters. Though a few still may be found in shallow Lake Weyba, they have almost disappeared from the rivers of south-east Queensland, including the Maroochy and the Mary. Chemical pollution is thought to be the cause.

The fishing was great, the old-timers say: 'You'd go out before breakfast and get as many fish as you wanted in no time, no trouble at all.'

George Burgess remembers when the seas from Caloundra up to Fraser Island would be black with travelling schools of mullet, all heading north to their different spawning grounds during the winter season. They would be netted off the beaches, and in the rivers and lakes, in thousands of tonnes. In those days there was no 'kerosene taint', the mysterious after-taste that has deterred many Brisbane people from buying this excellent table fish. It is thought to have something to do with the discharges from petrol refineries in the Brisbane River, or just spill-pollution from small craft. Mullet are an oily fish and could absorb oil from the water.

'There were hardygut mullet in thousands in the river,' said Burgess, 'but gill-nets mean dead fish. And there were no close seasons. Then big launches with refrigeration came from the south, working in shifts twenty-four hours a day. They prawned out the lakes; the mullet were scared from their old spawning grounds . . . .'

The prawn fishery, worked by local fishermen, was supposed to be worth a million dollars a year. The last bonanza years were in 1968–69, when all day long the waters of the biggest lake were churned into mud by circling trawlers. The prawns moved into the deep river, but the trawlers followed them there. As they got scarcer, competition became fierce. Some undersized prawns were allowed to go down to the Brisbane market, until it refused to take any more.

The fishermen then had to go out into Laguna Bay, crossing the tricky bar with its moving sandbanks when they could, to get bay prawns. Today, even these seem to have disappeared from the bay, and the bar, during training works at the mouth, became unnavigable except at full high tide.

Ten years ago you could buy school prawns, the sweet small prawns unique to the Noosa lakes, for twenty cents a pound, cooked, or green prawns fresh from the nets for bait. Now the only prawns are packaged, frozen ones from Brisbane, or fresh king prawns from Mooloolaba.

The lake fishing has never recovered. Numerous small bream, flathead and sole died in the prawn nets before they could be sorted. The development of cattle-breeding properties such as Cootharaba Downs and Elanda Plains — neither of which succeeded because of a mysterious disease called 'wallum ill-thrift' — probably had something to do with the decline. Both were in the Noosa River catchment area. At Elanda, crop-dusting planes spread weedicides and superphosphates, some of which drifted directly into Lake Cootharaba. With an extremely wet season following, these pollutants, plus pesticides from dips, were washed into the river system.

Lake Weyba, the nearest of the chain of lakes to the sea, rises and falls with the tide. It is very shallow, and therefore sensitive to pollution. It is a fish habitat reserve and has always been a breeding place for sea mullet. There are two fish habitat reserves in the estuary and lake waters. There are no fish sanctuaries, but trawlers are banned in Lake Weyba. In Weyba Creek, the big sea mullet used to school in the shallows under the old humpy bridge (a favourite fishing spot for line fishermen) where they were safe from nets. Fierce yellow-tail kingfish would congregate there to feed on fingerling whiting making upstream, and on bream, which in turn fed on the oysters clustered on the piles.

Today, Weyba Creek is no longer clear and blue-green as the sea. The creek esplanade has been developed for housing, with consequent run-off pollution, and silt has drifted up from the junction with the Noosa River as far as Weyba Lake from the sand-pumping operations at the mouth.

The new bridge was shortened with land-fill on either side — a cheap-jack engineering trick — so that the tides do not scour through as they used to do. It is now rare to spot a fish from the old bridge, and all but the hopeful few have given up fishing from the shady bank beneath its wooden arch.

The *Noosa Advocate and Cooroora Advertiser*, printed in Pomona, recorded in 1917 that the sewage situation in Tewantin was in a very bad, even a disgusting state. The overseer of the Noosa Shire Council complained that some owners had not emptied their backyard dunnies for six months, and 'it cannot be healthy for potential visitors'. It was 1976, nearly sixty years later, before deep drainage came to Tewantin. The first stage is now complete, with a pumping station near Doonella Bridge. A second pumping station,

Tewantin No. 2, began operation in 1979.

All sewage is now pumped over Cooloola Hill and then goes by gravity feed to a treatment farm near the council dump, well away from the sea. It is treated, dried, and then buried.

The whole of Noosa Heads is now sewered except for the highest sandhills, where septic tanks work well in a 120 metre depth of sand. In the clayey, swampy soil of Tewantin and Noosaville, septic tanks had to be pumped out and any overflow was likely to end up in the river. There is still a 'night cart' emptying backyard dunnies.

Deep drainage came first to Noosa Heads, where it was a necessity for the multiple-dwelling units around Little Cove and the national park road. The engineers' original plan allowed the treated effluent to flow into Weyba Creek, a fish-breeding reserve, and from there into the river and the sea. There was such an outcry from residents, more than a thousand signatures being put to a petition in Division 4, that the site of the 'temporary' works was moved to near where the first cottage of the Hawley family overlooked the river mouth. Sewage was pumped up the hill and, after treatment, flowed straight down again into the backwaters of Noosa Sound, from where it could be carried up river with every incoming tide.

One indirect result of the eutrophication of the water with nutrients was an increase in bream and blackfish (luderick) numbers. Green 'sea lettuce' began to grow on the rocks of the new development, and even on the rock wall backing Noosa beach. This attracted blackfish. Bream are notorious scavengers, and they also like rocks. The effect on the juvenile prawn population is not known.

Although treated for bacteria, the effluent allowed detergent, insecticides and garden wastes to go through. When the electricity supply broke down in the middle of the night (as happened sometimes during storms and bushfires) untreated sewage went straight into the river, to be carried upstream by the tides.

In the summer of 1975, the Noosa River had to be closed to swimmers by local health authorities, as the count of the bacillus *E. coli* was dangerously high. This may have had some connection with the pumping out of numerous stagnant Noosaville drains, in preparation for the introduction of deep drainage.

When septic tanks slowly began to replace dunnies, it meant less work for the council night cart, but more seepage of wastes into the river system. Even houseboats plying on the river were suspect, and now have to use closed chemical toilets.

But the greatest disruption to fishing and fish-breeding was undoubtedly caused by the building of Noosa Sound on what was Hay's Island: a series of winding channels, sand-flats and mangroves, which filtered the waters, slowed the scouring action of the tides, and made a home for marine life.

Even before Noosa Sound was mooted in 1972, a misguided 'fogging' programme was carried out each year by volunteer labour, to get rid of the sandflies, which were Noosa's greatest curse — the serpent of its Eden. Insecticides were sprayed liberally, in fine droplet form, over the kilometres of sandy flats exposed at low tide, where the 'biting midges', to give them their correct name, were supposed to breed in minute crab holes. What happened to the miniscule crabs, and the small fish that used to feed on *them,* was not considered. The midge problem was moderated, and the fishing fell away. Perhaps midges were an essential part of the food chain. And up on the hill, if you dig or disturb the sand at dawn or sunset, they are as bad as ever.

Whatever the cause — the faster tidal run, the sand disturbance and destruction of the sandy shallows by the swifter currents, the detritus stirred up by sand-pumping covering the seagrass beds, the insecticides in the water, the draining of mangrove shallows — all who have known the Noosa estuary over the years, agree that the fishing has deteriorated.

They don't need experts' reports, like the recently released Coastal Management Investigation study, to tell them this. The mangroves are gone, and so are the sandy tidal flats where the soldier crabs, marching in formation, would make a distinct rustling noise as they all turned together. Now there are barren 'grassed allotments' and riverside playgrounds and unit sites, ringed with rocks. The sandy shallows, where the fingerling whiting and flathead used to breed, are no more. The mud crabs have lost their habitat.

In the 1920s and '30s, they say, it was scarcely necessary to bait a hook. The fish were so thick in the backwaters at Noosa Heads that if you dropped in a line you could not miss. A quick pass of a net would haul in enough to feed the one hundred and twenty guests at Laguna House for breakfast.

As for the climate — well, climate all over the world has been strange lately, and Noosa was no exception. With a summer rainfall of over 2540 mm in 1976 (the average is 1701 mm) the wet weather extended well into the usually dry and sunny winter months. In 1977, on the other hand, it was extremely dry and the usual wet season did

A fine catch of sea mullet almost as big as the young fisherman. They were once netted in thousands off the beaches on their annual migration to spawn in the river and lakes. No 'kerosene' taint in those days!

not arrive. Again, in early 1978 and in 1980–81, Noosa Shire was declared a drought area, and there was severe rationing of water for gardens.

Although the Noosa area has become more 'fashionable' in recent years, family parties still flock here for their holidays, for there are still a few camping sites left among the posh new motels and glamour restaurants. Cars stream north from Brisbane at Easter and on long weekends, and every one that doesn't have a surfboard on the roof has a fishing rod alongside, or a boat and trailer behind. The fish population is under increasing pressure. How much longer will the Noosa River and its neighbouring beaches be renowned for their line fishing, when professional fishing has fallen off to such an extent that the local fish shops stock mostly imported New Zealand fish and frozen North Sea cod?

Fishermen say that the new bar and relocation of the river mouth have devastated the Noosa fishing industry, which is no longer viable. They say fishing in the big lake and in Lake Weyba deteriorated steadily over the last eight years, with a drop in water levels and the loss of all prawn and most mullet populations. Trawlers then went into Laguna Bay, but the channel has become so silted that all but one have left, and now operate out of Mooloolaba or Tin Can Bay. The final blow was the Queensland Fish Board's announcement that it intended closing down its Tewantin depot, where trading and freezing facilities had already been restricted since mid-1981.

Fishermen would like to see the rivermouth dredged and protected by two rock walls to break the force of the swell coming round the headland. 'At least when the bar was down near the Woods, you could still get out when there was a fair sea running,' says one. And whatever the experts in the Department of Harbours and Marine say, the fishermen are convinced that canal developments, dredging and pumping, and reclamation of wetlands have led to the industry's dramatic decline. Yet in 1981–82 amateur line-fishermen had a good summer season, with big flathead and whiting washed down the river by the fresh, after the first heavy rains for five years.

I have a 1965 *Fishing Guide for Australia*, which states: 'The beaches north and south of the Noosa River provide some of the best fishing in Australia.' This is no longer true of the southern beaches, at least, and more and more fishermen are going further afield to the seaward side of Fraser Island.

Dr Arthur Harrold, founder of the Noosa Parks Association and chairman of the Cooloola Committee, says that the river is deteriorating because of over-fishing, alteration in the river's flow and destruction of mangroves, sand-bars, and seagrass beds, partly caused by the development of Hay's Island. The beaches are also deteriorating under the pressure of people. The association has fought developers and managed to get the foreshores of most of the upper lakes, and some of the river bank, protected.

Present-day residents of the Noosa area do not realize how much they owe this dedicated band, working on a shoestring budget with voluntary help. Through their tenacious efforts, and those of Dr Harrold in particular, the Noosa National park with its remarkable mixture of rainforest, palm woods, open wallum and seaside heathlands has been saved for posterity. More recently they have fought sand-mining on the North Shore, peat-mining on the wildflower plain, and the Noosa Waters canal development proposal at Noosaville, and they have persuaded the Forestry Department not to plant slash pines in the western catchment of the Noosa River. This has added a most important section to the Cooloola National Park, the fight for which (coordinated by the Noosa Parks Association in Noosa Heads and debated in the Gympie Mining Warden's court) became a national crusade and added the name 'Cooloola' to the Australian consciousness, along with Lake Pedder and the Franklin River in Tasmania.

Dr Harrold has recently been recognized for his work in conservation by being awarded an honorary life membership of the Australian Conservation Foundation. The Noosa Parks Association is now keeping watch on developments on the north shore of the Noosa River, with the newly formed Noosa River Protection Committee to monitor proposals for canals and marinas.

7

Noosa beach: a priceless asset lost

It must be remembered that Noosa beach was not destroyed by fire, flood, storm, tempest or other act of God. It was destroyed by the Noosa Shire Council and certain government departments . . . by the most effective method known to modern man, in the building of a boulder sea wall.

Max Walker, F.A.S.A., long-time resident of Noosa Heads, in a letter to *Noosa News*, 1978

The once clean and beautiful curve of golden sand that formed the beach at Noosa Heads has come and gone with the years. Geologists say that the Sunshine Coast beaches are about five thousand years old, and in that time they must have been scoured many times to the underlying rock in cyclonic storms.

The same natural forces that built up the beach at the beginning have always renewed it after it has been scoured away. It is only in the last hundred years of white settlement that the beach has eroded to the point where sea was breaking on the esplanade wall at high tide, with no beach left and no buffer sand dunes behind it.

This can be blamed partly on unwise development, such as the building of houses and car parks on top of the frontal dunes; but it also seems certain that either the land is gradually sinking, or the sea is rising. This is demonstrated when one looks at early photographs of the large casuarinas that once lined Noosa beach or grew below the present national park road. Older residents say that these trees have disappeared in the last fifty years.

In the summer of 1967–68, several disastrous cyclones either struck nearby or passed on a parallel course out to sea, whipping up huge waves. The sand was once more scoured away, as it had been in 1931 and 1947, exposing large pebbles and underlying 'coffee rock' (a dark, coffee-coloured indurated sandstone, containing much oily peat and soft enough to scratch with a fingernail). Property owners on the beachfront, seeing the remaining sand being eaten away in great gulps with each spring tide, began a panic dumping of rock fill in front of their properties, on the beach itself. They did not wait for the natural, cyclic build-up of sand. In the last two years without cyclones, 1977–78, sand came back along

The lagoon, Noosa Woods and Hay's Island, 1947. The beach was then a 'clean and beautiful curve of golden sand'.

Noosa Beach from the new national park road, 1945. Beach House can be seen in the middle distance, Mt Tinbeerwah on the skyline. Casuarinas shade the beach.

*Before the rock wall was even thought of: the natural
curve of Noosa Beach, 1938. (By courtesy Beth Smart.)*

Thirty years later. The beginning of the end: the scoured beach after the January 1968 floods and high tides. Rocks have been dumped in front of the new Noosa Court Motel. Indiscriminate rock-dumping by beachfront owners attempting to protect their properties has caused more scour. There are sharp boulders in the sea and underlying rocks have been exposed. (By courtesy Kevin Freeman.)

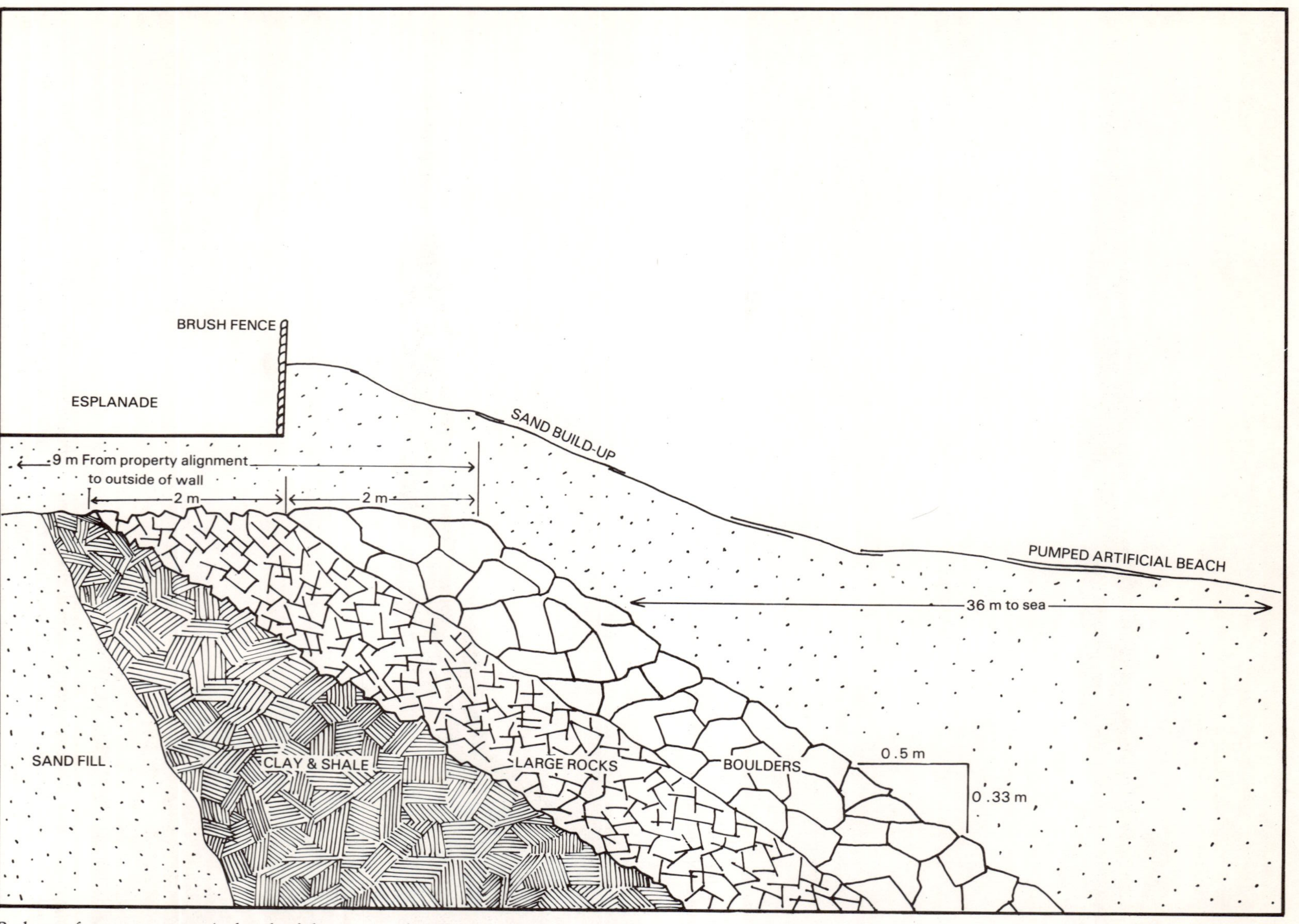

Redrawn from an engineer's sketch of the proposed wall (1968), showing rocks completely buried in sand.

the coast in vast quantities — but not enough to cover the rocks at Noosa Heads.

The reason the beach could not regenerate, as hydrographers and geomorphologists agreed, was that the steep-angled boulder wall had altered the natural slope and curve of the bay.* This wall was built in 1968–69 by the shire council on the advice of its consulting engineers and with the blessing of the Department of Harbours and Marine.

Urged by property owners on the seafront who wanted their land and buildings protected, the council called for tenders for building a rock wall to stretch the whole length of the beach (820 metres). This was later extended to protect the Noosa Woods parking area, at a total cost of more than $200 000. The whole shire is still paying extra rates for it, while residents of Hastings Street pay a higher, 'benefited persons' rate.

The wall did nothing to protect the *beach*, only the property behind it. In a press statement in 1968, the Shire Chairman, Councillor Macdonald, said he was personally in favour of the wall, 'which will be several hundred yards long. On completion, a sand pumping dredge will start to fill in behind and cover the wall, *thus creating new sand dunes and a new beach*[author's italics]. It is estimated that 60 000 cubic yards of sand would have to be pumped from the Noosa River to the beach.'

(In fact, an artificial beach three and a half metres high, thirty-six metres deep and 820 metres long would have required about 72 000 cubic yards, or 90 000 if the area behind the wall was to be raised as an artificial dune. The most that was ever pumped at one time was 3000 cubic yards.)

The consulting engineers, John Wilson and Partners, issued a neat diagram to show 'the project as envisaged'. There would be a three and a half metre high rock wall with its base at low-water mark; the area behind would be filled with sand, and sand would also be put on top of the wall and poured out on to the beach in front. A sand-pump would pump new sandhills over the rocks, and 'the wall will no more be seen'.

---

*'The succession of cyclones in the late '60s were unusually severe. Although we shall get similar conditions again, it will be at very rare intervals . . . . This is a natural hazard . . . .

'The alignment of the beach at Noosa Heads is quite different from that of the Gold Coast beaches, and Noosa would have had to wait longer — for the appropriate periods of constructional waves — for the beach to recover. Instead of showing patience, a sea wall was built which immediately wrecked the beach.'
  Professor T. Langford-Smith, Professor of Geography, University of Sydney.

This was a sad joke. For a while, some black filth was pumped from the backwaters behind Hastings Street: the detritus of swamps, the overflow from motel sumps, mangled fish and rotting vegetable matter, mixed with some grey sand.

This was washed away with the next heavy rain, and the once-pellucid waters of Laguna Bay managed to clear themselves. The pump was found too expensive to run, after pipes had been extended down to the sandy channels beyond the Woods to get clean sand. The pump was leased out to a developer to bring in revenue. Millions of tonnes of sand, instead of going on to the beach, were used to raise part of the natural flood plain in the river mouth, the maze of mangroves, sandy islets and winding channels that was Hay's Island.

The rock wall began to subside the following year. Large boulders lay hidden in the surf, and a visiting swimmer, body-surfing on the beach, crashed head on into one and was killed. Still nothing was done about covering the rocks, though steps were built down over the wall in two places for elderly people and children who could no longer make their way to the beach in safety. Both sets of steps have since been washed away twice. In the next wet season, loads of red clay and small, sharp pieces of shale rock were dumped behind the wall, and sand was spread on the 'esplanade' thus made above it. Clay and sharp rocks disfigured the beach for a year or two, and the sea was discoloured after heavy rain; but gradually both were buried by the all-forgiving sand.

The waves still broke in their classical pattern round the Point but, as they reached the foreshore at high tide, they came up against the rigid straight line of the wall. Instead of breaking smoothly on a slope of sand, leaving any loose sand they might be carrying as they retreated, they smashed against the rocks and rebounded. This set up a surging motion as they met the next incoming waves. Instead of being spread on the beach, sand held in suspension was dropped in a bar further out (see diagrams, p.78).

Another factor which was not understood was that the effect of trapped air, compressed between the falling weight of the waves and rocks, was to increase greatly the pressure per square metre of a given weight of water. This tends to break up the wall, and is a danger to any swimmer caught between rocks and wave. Professor Gordon McKay, who retired from the University of Queensland in 1978, had warned at a public meeting in Noosa Heads that enormous quantities of sand would be needed — far more than had been suggested.

'I don't like rock walls,' he said. 'The wall is a property protection . . . . If you want a beach as well you have got to start bringing large quantities of sand.'

The engineers' plan provided for an artificial dune on top of and behind the wall '. . . as a sand reservoir in rough weather. The dunes are to be planted with suitable vegetation and watered till well established.'

None of this eventuated. There are now only two casuarinas left of those planted in front of the car park. The row of pines, planted behind these for protection, soon withered and died in the salt. A short length of brush fence and a metre or so of sand were placed in front of the car park. Elsewhere, the area above the rocks consisted of eroding clay, and this was washed down into the sea with every rain. Sea lettuce grew in a green fur on the big boulders. There was no access to the beach except by clambering painfully from rock to rock.

Apparently, after the wall was built and the sand-pump had been bought at a cost of around $20 000, there was no money left for running the pump, paying council workmen to man it, and moving the dredge to new sites off Noosa Woods. Instead, a wooden groyne was placed a few hundred metres along the beach, and for a while did trap some sand, while the beach beyond became deeply scoured. Then a storm smashed part of the groyne so that, while it remained an eyesore and an obstruction, it did nothing to arrest lateral sand movement. 'Though the top of the groyne could be exposed in rough weather', read the engineers' optimistic proclamation, 'it should be buried by the new beach and not detract from aesthetics.'(!)

Before the 1947 cyclone: the beach below the lifesavers' club, 1941. (By courtesy Kevin Freeman.)

Thirty years later: Noosa beach at high tide in January 1971. Rocks instead of sand. (By courtesy A. G. Harrold.)

The abortive groyne on Noosa beach. It was never kept in repair, became an eyesore and was finally removed in 1977. Before being holed, it did cause a build-up of sand on the south-eastern side but some scouring beyond. Note rock wall in foreground.

There was also a recommendation to council that future building permits should be rigorously controlled between Hastings Street and the beach. 'Under no circumstances should builders and owners shift any sand.'

In 1970, Professor T. Langford-Smith, then Professor of Geography at the University of Sydney, came to Noosa Heads for a holiday. He had been a regular visitor for years, and intended to settle in this beautiful area eventually. He was horrified to see the ugly boulder wall that now defaced his favourite beach.

He wrote in protest to the Queensland government, sold his block of land, and has never come back. In a letter to the Coordinator General's Department, dated 12 June 1970 (the year after the wall was completed), he said that, from his knowledge of the locality, his experience in coastal research and his inspection of the rock wall, he was dismayed to see such a magnificent beach wrecked'.

His suggestion was that owners of beachfront properties should be forced to surrender the front (seaward end) of their blocks — some had even built swimming pools where the beach used to be — and the present sea wall should be removed,

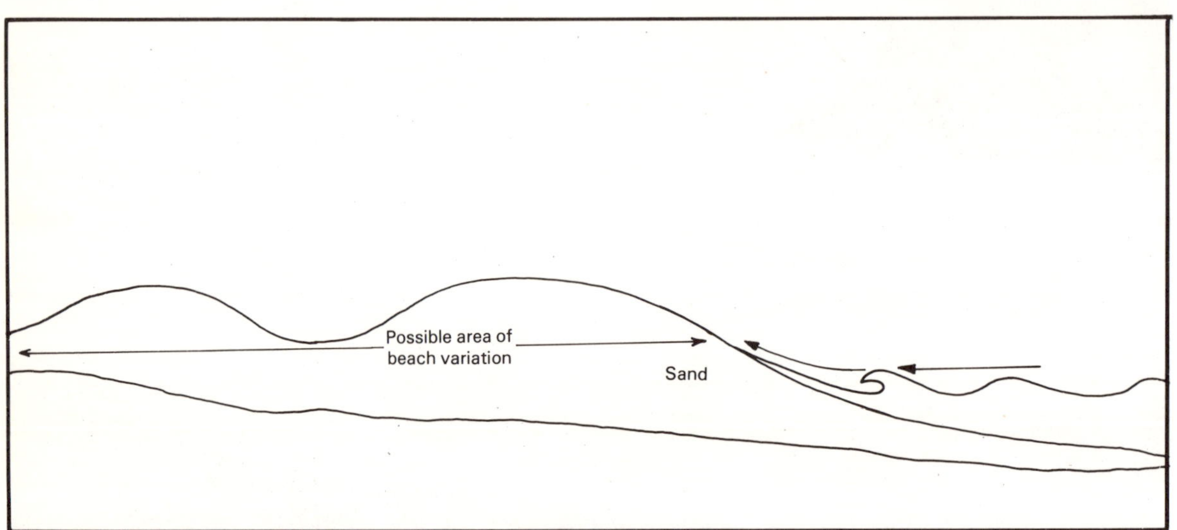

A. Dune formation at Noosa Heads.

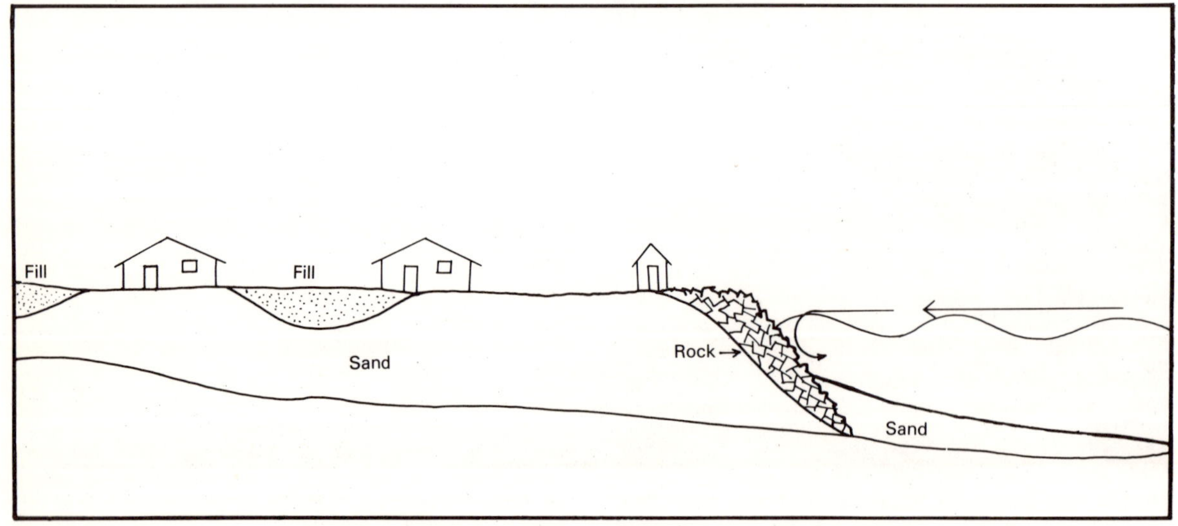

B. The erection of houses and rock walls causes scour and wave surge.

After some heavy (but not unduly rough) weather, the new beach developed a three metre cliff of sand. Warnings of the sudden drop had to be put up at night and a *dangerous sea developed at high tide for swimmers. This picture was taken in January 1979. By the end of 1981 no sand was left at all. (By courtesy* Noosa News.*)*

while perhaps a very gently-sloping wall could be put further back to protect buildings until the beach had a chance to re-establish itself.

His letter was all but ignored. The newly-constituted Beach Protection Authority replied shortly and sharply that '. . . in the absence of detailed knowledge of the actual erosion processes . . . the provision of a boulder wall certainly offers the best chance of protecting the existing property. There has never been any suggestion that a beach would necessarily re-establish itself naturally in front of the wall.

'To remedy this "lack of beach" situation, it was intended to pump sand from the Noosa River . . . . Difficulties however arose which limited the amounts of sand which it has been possible to pump to date, but additional quantities will be pumped in future.' (They were not.)

In an interview in Sydney, the Professor said he was not up to date on recent developments at Noosa Heads, but he did not hold out much hope for Professor McKay's plan, developed by the Beach Protection Authority and carried out in 1978. He said it was trying to force the sea back, instead of yielding a bit of land in order to gain space for a natural build-up. Professor Langford-Smith said:

It is ridiculous, in geological terms, to say that the beach 'has not come back' after three or four years. The present Noosa Beach is five to six thousand years old — the length of time since the sea-level last rose — and a few years is nothing in that time-scale. The beach has been eroded down to underlying rock, and returned again, countless times; it has healed itself in cyclone-free years. Sand-pumping has helped the Gold Coast beaches to recover, but that was only the catalyst — the sand was

79

lying offshore ready to return when conditions were right.

The steep, straight wall, which was below high-tide mark, is both dangerous and useless. It sets up a surge which makes the natural build-up of sandhills impossible.

And all this was already known. The University of New South Wales has experts who have studied these factors for many years and are ready to give the benefit of their advice. But it is not asked for, nor taken.

In fact it was known forty years earlier, when there had been a similar move to pile rubble on the beach to save it from erosion. A band of local residents formed a Beach Progress Association and got expert advice that such action would save property, but destroy the surfing beach. So it was not done, and in time the sand returned.

The Noosa Beach Progress Association was formed in about 1927, before there was even a road to Noosa Heads. The members collected funds (the 'Beach Improvement Fund'), put up swings for children, and erected the first shelter shed on the beach (twice moved back as the sea advanced). The original fibro-roofed shelter is now situated at the rear of the parking area, beside the road.

Early photographs from Laguna Hill show a thick mass of trees on the seaward side of Hastings Street for its whole length, except for the sand-blow opposite Laguna House. This open space is

Back to square one: the rock wall re-exposed after moderately rough seas and high tides, 1981. Even the narrow 'esplanade' of sand at the top is gone, and the rocks are falling into the sea. (By courtesy Noosa News.)

An eagle's view of the village street, still thickly shaded in the 1940s. (Photo by Murray Views of Gympie.)

The beautiful greenery on Hay's Island has all been swept away, 1970. (Photograph by Bernard Kuskopf.)

For comparison: the main street thirty-five years earlier. The fragile sandspit carrying the main street was then thickly wooded, as was Hay's Island (left). (By courtesy B. Lintott.)

now the site of a new, thirty-unit multiple dwelling, set down into what was once the frontal dune. Near it is Tingirana Arcade where, in spite of the engineers' recommendations about seafront building, the dune was removed completely to give a level walkway to the beach. The hundreds of tonnes of good, yellow beach sand taken to make room for it were not spread over the rock wall, though offered to the council for sixty cents a tonne: instead it was sold and carted away to a building site at Sunshine Beach, where there is plenty of sand!

Most of the strip of rainforest and salt-resistant dune vegetation that formed a windbreak from the sea has now been removed to make room for modern motels with swimming pools and lawns. Some of the last to be cleared was beside Paramount Apartments, at the far end. At the old Seaview, and at Kelly's place, some of the former rainforest can be seen.

According to old-timers, the local Aborigines said that the mouth of the Noosa used to come out right under the headland, nearly half a kilometre south-east of its present outlet. Frank Crank remembers the year when it looked as if this might happen again, with the sea 'lapping the top' of the frontal dunes at the car park. Twenty men, he said, were standing by to evacuate people if the sea broke through. One of the residents on the front, Bill Lowe, got such a fright that he left for Cooroy and built in the safety of the hills.

George Burgess remembers the 1893 floods which inundated Brisbane. At that time it rained for nine days at Noosa and blew a hurricane. The lakes rose by nearly two and a half metres, and cattle and horses were feeding on islands surrounded by floodwaters.

Mrs Olive Freeman remembered a trickle of seawater from a huge wave reaching to where her front gate is now, at the corner of Morwong Drive. And that was before the buffer sandhills had been lowered to make the paved parking area. The freshwater lagoon (where Ocean Breeze units are now) was sometimes salt when a blow coincided with king tides, and the sea would back up across Hay's Island and into the lagoon.

In the 1931 flood, according to Howard Parkyn, waves were breaking one and a half metres high at Munna Point. Water ran across Hastings Street, and was a metre over the approach to the new bridge at Doonella Lake, cutting off Noosa Heads from Tewantin. The Parkyns' boat ran milk, bread and mail to the Heads, as the Noosa Coastal Highway had not then been built and the town

Floods at Tewantin, 1931: only the top of the arched bridge is still above water. Parkyn's boat shed is isolated. (By courtesy Frank Bickle.)

was completely isolated. (In January 1968, when the Noosa area had 760 millimetres of rain in two days, the river was coming down at 10 knots, eroding the sandspit at the Woods, and they were catching mullet in the main street at Noosaville. The bridge was once more out, and the Lakes Motor Inn, built on a former mangrove swamp, was flooded.)

The late Bill Ross, Edgar Johns and many others said that in their memory the beach went right out as far as First Point. At low tide you had to walk almost to the first headland for a swim. In the corner where Los Nidos is now, there used to be some very big native pines, and sand shaded with banksias curved below where the rocks are now. Bill Ross recalled:

When I knew it, the beach was about three hundred yards out. Where the kids used to picnic is now under water, even at low tide. The sandhills came right down to where the water is now, covered with she-oaks and grass. The cyclones just took those dunes away, millions of tons of sand. When the first bad blow came about forty years ago [i.e. 1931] someone came and told me, 'Beach House is in trouble.' The water was only about six to eight feet from the front veranda. 'If this keeps up', I said, 'she'll be gone in the morning.'

Beach House was the first boarding house to be established on the frontal dune and it is still there, though the inside of the old weatherboard has been relined and refurbished as Annabelle's Restaurant. The first shop on the beach itself was a kiosk run by Walter Hay and his descendants, the 'Shanty', which sold soda fountain drinks. The Hays also put up a square wooden bathing box, later used by the public as a dressing shed. It was eventually washed away.

Walter Hay had removed the small house above Bayview across to Hay's Island, where he cleared some of the big trees for his market garden. Meanwhile the old Bayview crumbled away with white ants. It was pulled down, and the site was bought by Johnnie Jones whose new boarding house was called Hillcrest (now Halse Lodge) — rather a misnomer as the hill towers above it.

In 1926 Charlie Freeman bought the old Portion 171, high on the hill. He was an excellent fruit farmer, established a banana plantation and a citrus orchard on the sunny slope, and grew mangoes and pawpaws. His farmhouse, Noosa Vale (now Hack's), had a sweeping, panoramic view of bay and estuary. The garden was gay with bougainvillea and bignonia in clashing orange and magenta against the green of three tall palms. He built his own road and bridge to the house.

Noosa: the lagoon in flood, 1931. (By courtesy Olive Freeman.)

*Looking towards the east in the 1930s. The sea is right
out beyond First Point.*

*The road to Freeman's farm, 1920s. To the left of the
Ford T buckboard is Bayview.*

Betty Freeman, Hay's great-great-grand-daughter, with some of the first banana crop from Laguna Hill. Pineapples are growing in the background. (By courtesy Olive Freeman.)

Charlie Freeman's farmhouse, below Laguna Lookout. Note the flagpole. (By courtesy Olive Freeman.)

Kevin Freeman, Hay's great-great-grandson, on the deck of his tourist launch, Miss Laguna II, circa 1948. (By courtesy Olive Freeman.)

His children, Kevin and Betty, had live koalas to play with. They also had a wide, safe beach shaded with casuarinas and banksias in which rainbow lorikeets fed and squabbled.

The ancestors of the whistling kites that still nest in the three big blue gums below Halse Lodge (the former Hillcrest) sometimes dropped a small bough on the roof as they flew over. By now the Donovans had moved to the Royal Mail at Tewantin, and Mrs Donovan's relatives took over Laguna House, keeping up the tradition of beautifully cooked fish meals and carefully packed picnic lunches.

There was often excitement when ships in trouble sheltered behind the Noosa Headland. One, remembered by Lionel Donovan, was the *Antonio Pederice*, an Italian four-masted barque.

Toy bears were not needed for Noosa Heads children in the early 1920s: a live koala plays with a child outside Bayview. (By courtesy Olive Freeman.)

The Massoud brothers, George (left) and Bill. Bill holds his citation from the Royal Humane Society for a boat rescue in Laguna Bay. George points to one of the early entries in his father's account books. (By courtesy Noosa News.)

In the cyclone of 1910 she lost her rudder, but managed to anchor and send a rowing-boat in with a message. She had no wireless. John Donovan rang Brisbane on his little 'earth circuit' line, and asked for a tug to be sent to her rescue.

In a later blow, a boat sent up a distress flare in the bay, but no one dared go out to her. The bar was quite impassable. The story is that young Olive Freeman (wanting perhaps to be a modern Grace Darling) began pulling her rowing boat round from the Woods lagoon by a rope, preparing to launch it in the more sheltered corner of the beach opposite her home.

Not to be outdone by a mere girl, Bill Massoud ordered her back and launched his own boat from the beach. He found the men, with a swamped engine, too seasick to save themselves from being dashed on the rocks. He was able to take them in tow and bring them safely back to the beach, for which exploit he received a citation for bravery from the Royal Humane Society.

All the brothers Massoud made a team to provide the area with a volunteer rescue squad. Under the command of Mr Massoud senior, the boys went out in 1943 on an unsuccessful search for the *Werita*, which disappeared with all her crew of five. (Bill was in New Guinea at the time with the R.A.A.F. air-sea rescue squad.)

In the 1920s the young Harry Johns came down from Restdown, where the orchard was not doing very well, to manage a store at the Noosa River mouth for his wife's family, the Marsdens. In the summer of 1928–29 there was a very bad cyclone which passed near Noosa Heads. The word went round that Harry Johns' home and store were in danger.

A working bee of fifty men was organized, and brought down from Tewantin and Noosaville by Jack Parkyn in the *Miss Tewantin I*. By then the tide was rising, with a cyclonic surge which was cutting away the sandhills and their protecting casuarinas, while waves were washing round the stumps of the house and under the wooden veranda. Even a draught horse was called in to help. Men prised out nails and removed whole sections, the horse pulling, and in a few hours the whole place had been demolished and carted bodily away from the hungry sea. The old stumps have been covered by sand and re-exposed three times since then, as the mouth moved back and forth.

It was set up again further along Hastings Street, still on the seafront, as Seaview Flats. Along the front, in 1930, there were seafront flats belonging to Edgar Johns, Noosa Heads Lifesavers' hut and tower, two dressing and shelter sheds, set on stilts on the sandhills, Bill Hayden's cafe, Harry John's house and his Seabrae Cafe, the back of which opened on Hastings Street and was the post office and store, with the town's only petrol bowser. The Johns family put stakes of paperbark trees at the foot of the sandhills, to break the force of the waves while letting the sand wash back through them.

Laguna House was managed for a while by Listers and Walters, then by the Gibneys, before it was bought by Thatchers (who owned Thatcher's Caravan Park). The next owners were Gilchrists, Smarts, and Pooles. Mrs Smart's daughter Delse Poole gave the old building to the Lutheran Church; it was carted away about 1961, and set up on top of a hill at Coolum where it is still used as part of a holiday camp.

By 1940, Noosa Heads beach was famous far afield. Aeroplanes landed on its wide expanse to give joy-rides. 'Sandgarden' competitions were organized at weekends by the Brisbane *Courier*. The 'Vita-sun Oil' man, John Paterson, was a regular visitor to Laguna House with his muttonbird oil for sunbathers. He had a stuffed muttonbird mounted on his pith helmet, and carried a spray-gun with which he sprayed the beachgoers with his mutton-bird suntan oil. Another muttonbird adorned his Rolls Royce. Another regular at

During the 1928 cyclonic surge, Marsden's house and store (occupied by Harry Johns) was threatened by the rising sea. More than fifty men came down by boat from Tewantin and removed the house. (By courtesy Olive Freeman.)

The Miss Tewantins: Top: Miss Tewantin I *at Parkyn Bros Wharf. (By courtesy Frank Bickle.)*   Bottom: Miss Tewantin II *cruises down the river. (By courtesy Howard Parkyn.)*

The Sea Eagle made a forced landing on the wide expanse of beach, then proceeded to give joy-rides to those game enough to go up in her.

Panoramic view of Noosa Heads beachfront in the early 1940s.

E. Johns' Flats   Freeman's Farm   Laguna Hill   Lifesavers tower and clubhouse   dressing shed   dressing shed   Hay ca

G-AUGH, the Wunda, on Noosa Beach, taking passengers at ten shillings a time.

sandblow
(now car park)

paperbark fence
in front of C. Johns'
(now Barry's Bistro)

Harry Johns'

Seabrae Cafe (H. Johns)

Sand-garden competition run by the Courier Mail *in the 1930s — a regular feature of the wide, sandy beach. (By courtesy Frank Bickle.)*

Schoolgirl winner of the 1936 sand-garden competition. (By courtesy Olive Freeman.)

Laguna House was 'Pop' Calvert from Glen Applin homestead near Stanthorpe, in his smart Jaguar sedan.

To tourists of the last generation, a holiday at Noosa Heads was synonymous with Laguna House. A popular song of the early '30s, 'She's the Lily of Laguna', seemed particularly apt. Even if they stayed somewhere else, to most people Laguna House, halfway along the little main street, was the centre of life and gaiety.

Hillcrest up on the hill was more staid, more for family parties and children; so was Muller's Beach House on the seafront, managed by Bert Groom who had married a Mrs Hay, a widow. But Laguna House, where people changed to have dinner in the big dining room which seated one hundred and twenty guests, and where the table gleamed with starched white cloths and table napkins, silver cruets and crystal jugs, was a great place for romance, and for honeymooners. There was a nine-day scandal in the town when the proprietor of one guest house ran off with the wife of the owner of the rival guest house.

On New Year's Eve the Laguna House party was the event of the year. Coloured lights were strung along the forty-eight metre frontage, and the band played till two in the morning. In the warm summer night, if it was not raining, couples drifted out on to the wooden verandas or across the road to look at the sea by moonlight. By day, drying beach towels and costumes hung over the railings and made a bright splash of colour. Guests could sit on the high veranda and look across the sandblow at the sea.

For more than fifty years this landmark endured, the name painted on its iron roof in large letters visible from the Laguna (or Tingirana) Lookout. When John Donovan took it over from the Bainbridges in 1906, there was only a track in front of it and a samphire swamp behind.

'Pop' Calvert and his Jaguar outside Laguna House, circa 1947. (By courtesy Kevin Freeman.)

Photo by Murray Views shows Beach House, a dark,
two-storeyed building in the middle distance.

The dining room at Laguna House, resplendent with
white napery and crystal, 1950s. (By courtesy Mrs Beth
Smart.)

One of the last owners of Laguna House, Mrs Gilly Gilchrist (now Mrs Beth Smart of Maleny), dressed for the New Year's Eve ball in 1945. Coloured lights stretched the whole length of the buildings. (By courtesy Mrs Beth Smart.)

On its site is now an area of lawn and modern shops and cafes, the Laguna Shopping Arcade. Traces of the old, red-cement tennis courts, said to be the first electric-light courts in Queensland, can be seen at the western end.

Late in 1969 a Sydney firm, Wynyard Holdings, put up a notice of intention to build in Hastings Street, on the site of the old sand-blow which had never been built on and gave an alternative access to the beach. Objections to the plan were called for. They came in a deluge. Yet the council gave its approval.

The plan provided for the first high-rise building at Noosa Heads, something that would have changed the whole character of the little town which still managed to retain a 'village' atmosphere as part of its charm. The plan was for a nine-storey complex on the seafront, with walkway to the beach, shops, a conference room and motel.

There followed one of the bitterest fights for and against ever to divide the residents of Noosa Heads. At the instigation of Mrs Marjorie Harrold, the Noosa Planned Progress Committee of interested citizens was formed to fight the big developers from Sydney. Funds were collected, and the matter was taken to the Court of Local Government Appeal. Pending hearing of the case, work could not begin on the building. Meanwhile

1977: Not the devastation from a cyclone, but a new development by Wynyard Motels of Sydney. Site of the old sand-blow opposite the former Laguna House, it had never been built on. A four-storey home unit block has now been built after removal of the original sandhill. This was where a nine-storey building was planned and successfully fought by local residents.

the summer season came and went, with its usual cyclone scares, and the developers began to have second thoughts. They asked council to guarantee that the rock wall would protect their property. Council rightly refused any such guarantee, though they had given permission, in principle, for a high-rise building on the seafront.

However, when the case came on, the developers flew up the big guns from the south. Two Q.C.s, Professor Bunning (an eminent landscape planner), an army of engineers, draftsmen and other 'experts' arrived. Planes flew overhead photographing the site. Shadow boxes were constructed by both sides, to show the degree of shade falling across the beach in the afternoons. One advocate of high-rise suggested that tall buildings would be beneficial in saving people from skin cancer by shading the beach!

In the meantime an eighteen-storey building was mooted on the other side of Hastings Street, almost opposite. Residents were horrified; another Gold Coast was looming up.

Meanwhile, a town plan for Division IV, including Hastings Street, was in preparation, with an interim development by-law to control building. Mr Challinor, a highly qualified town planner, was appointed official town planner for the shire. His advice to the council in 1969 was to refuse the permit, as he considered ' . . . the proposed development not suited to the subject site and . . . high rise development is not desirable in this section of Noosa Heads . . . the proposal is not in accordance with planning proposals for Noosa Heads.'

In spite of his advice, approval was given in October 1969. Objection was immediately lodged with the Local Government Appeal Court by residents. Before the case came up in March 1970, council announced at a public meeting that Mr M. Poole of John Wilson and Partners had been appointed town planner in place of Mr Challinor. No explanation was given.

When the residents' appeal was heard, Mr Challinor's stand against high-rise buildings was rejected as evidence, as he was no longer town planner. Mr Poole, when called, stated on oath that he had not inspected the subject site, and had no opinion on the application. He also stated that there were, as yet, no planning proposals for Noosa Heads.

The appeal was dismissed by the Court. Less than a month later, Mr Poole announced in council that the town plan for Division IV at Noosa Heads was now complete!

It is difficult to understand how the most valuable area in the division should have been left till the very last in his planning, and then suddenly be completed in one month.

One of the strongest arguments against high-rise had been that of traffic congestion in the dead-end street, both during construction and as a result of greater population density.

Another was that the breathtaking view from the crest of Motel Hill, of the sapphire sweep of Laguna Bay with its backdrop of coloured sands, would be lost behind square stone boxes. It was feared that, once the precedent was established, the whole seafront would become lined with skyscrapers.

The effect of high buildings in funnelling strong sea winds was another question raised. Also, the plan involved excavating the natural sandhill for a car park which would have been almost below sea-level. Another reason against was the lack of fire-fighting equipment available for dealing with fires in high-rise buildings.

The case was heard in Brisbane, so that there was almost a mass exodus of local residents for the hearing, both those who were for and those who were against. The Court not only found in favour of developers and council, but, to teach private people that they had better not have the temerity to challenge big business in future, a day's costs (estimated at $2000 — and this was over ten years ago) were awarded against the appellants. The opposition's whole case, with lawyers' fees and witnesses' fees, had cost no more than this. They were given time to pay and, on appeal, the sum was reduced to $1500, Mrs Harrold as the nominal appellant bearing most of this cost.

Gloom prevailed among the Planned Progress group, but all was not lost. A high tide and a bad storm (not even a cyclone) having coincided, the developers let their building permit lapse, and the site remained open until 1977, when a four-storey building went up. It made no provision for a walkway to the beach.

The new town plan stated that 'high rise in this part of Noosa Heads is not desirable.' The town plan was formally implemented in 1972, and no further high-rise applications could be made. This effectively dealt with the eighteen-storey building as well.

Cyclones and tornadoes have taken their toll of the thick greenery that once was Noosa Woods. It is now an official camping area, and some trees have been cleared to make room for more sites, while two red-brick 'amenities blocks' stand out like sore thumbs. Electricity came to Noosa Heads about 1948, the year of one of the worst cyclones.

In a later blow, a camper at the Woods was killed as he groped his way out of his tent through the boughs of a fallen tree, and touched a live wire the tree had brought down.

In December, January, and February of 1947–48 there were three destructive cyclones that passed near by. Mrs Smart (formerly Mrs Gilly Gilchrist) remembers how Laguna House lost part of its kitchen roof, which landed over on the beach; and how campers who had been blown out of the Woods were brought in for shelter. One woman seemed quite calm, then she started to bawl: 'I've left my facewasher behind!' she wept. 'I haven't brought my facewasher!'

The new electric stove and the wood range were flooded, so they lit the wood copper in the laundry and boiled one hundred and fifty eggs for the guests' breakfast, making toast on the coals, and billy tea. There were five builders staying there at the time, and by that night they had the roof on again.

Mrs Gilchrist went over to Beach House and saw that the cyclonic rain was pelting through a hole in the roof. Downstairs in the kitchen she found the proprietor's wife mopping away at the water on the floor. She thought it had seeped in through the window and in the general noise and roar of wind and surf did not know the roof had gone.

In the next blow the undercut sandhills began to give way, and the children's swings went under water. The big shady casuarinas along the dunes toppled over, were swept away, and finished up in the river mouth. Then it was the turn of the buildings: the surf lifesavers' hut was moved back once more and the tower was rescued and laid on the ground further back, but the two wooden dressing sheds slid into the sea and were smashed to pieces. The beach was battered and scoured, and narrower than it used to be; yet, left alone, it re-established itself. The sand came back. Seedling casuarinas established themselves right along the spit beyond the Woods.

Then, in 1954, another three cyclones battered the beach, exposing coffee rock and the large pebbles that underly it at the eastern end. The summer of 1967–68 saw the beginning of the end, with heavy rains and Cyclone Dinah, the one that brought such destruction to Gold Coast beaches. That was when private individuals began dumping loads of rocks on the beach, including the owner of the new Noosa Court Motel. The beach deteriorated still further as a result, and at last the

The 1947–48 cyclones: **A.** *The lifesavers' tower, twice moved back since 1930, lies on its side above the sea. The children's swings are under water and houses are threatened. (By courtesy Kevin Freeman.)*

B. *The lifesavers' shelter has been removed. The men's and women's dressing sheds are falling into the sea. (By courtesy Kevin Freeman.)*

C. *The beginning of the end: three destructive cyclones in December, January and February sliced away millions of tonnes of sand from the dunes. Here the big, shady casuarinas are threatened.*

D. *Here they are going . . . waves smash into the foreshore and undermine the roots.*

E. *Going . . .*

(*Photos by Kevin Freeman.*)

F. Gone. Only the roots remain, sticking up in the foreground. Others were washed up in the mouth of the river. (By courtesy Denise Thatcher.)

desperate measure was taken of building the boulder wall.

For one whole year the beachfront was a mess. It was a wet year, and the beach was a mass of red clay, bulldozers, dirt-stained waters and jumbled rocks. But at last it was finished, and the wall withstood the next cyclone and did not collapse, as some had feared. Nor was it covered in sand, as the council had promised.

Cyclone Daisy came very close to Noosa Heads in 1972; in fact the eye passed over Laguna Bay. Another cyclone, not so bad, arrived at Easter: that was Emily. In July 1973 a winter cyclone did some damage. Then came Wanda in January 1974, the one that brought the disastrous Brisbane floods, raising levels in the Noosa River and eroding the piles of the new Noosa Sound Bridge. The sandy point at the Noosa Woods camping area, which started to go in 1968, was finally washed away. Rocks and rubble were tipped on to protect the Woods.

It was Cyclone David that gave the beach its longest battering. This cyclone passed right down the east coast without crossing, maintaining its intensity and whipping up great seas. This was in January 1976. At high tide there were fears that the sea might break into Hastings Street.

The beach steps disappeared again, large trees were washed down the river and flung up on the 'esplanade'. Rocks from the wall fell down on to the bathing beach and the clear waters of the bay were filled with brown and red clay from the hastily tipped-on reinforcements.

At the Noosa Woods a bad whirlpool formed. the river cut right in under the point, bringing deep and swift-flowing currents where rowing boats used to moor and children once learned to swim, or paddled safely in canoes. Only a row of sandbags saved places such as Edgewater Units, when the channel round Hay's Island rose to lap their verandas. At Munna Point and Noosaville, ground-floor houses and shops had water flowing through them. The Lakes Motel was once more flooded out. Some time after the storm had passed, a small boy bathing at Munna Point swam round the end of a moored boat and disappeared, apparently in a whirlpool, and was swept out to sea. His body has never been recovered.

All this was caused by a tidal flow rather than heavy rain, as there was very little rainfall associated with this cyclone. Many people blamed not just a natural disaster, but a man-made one, completed in December 1973 but begun nearly two years earlier: the canal estate in the river mouth, Noosa Sound.

Noosa Shire Council Report on Erosion Prevention at Noosa Beach 11 July 1967

When Noosa beach was surveyed, a public road and esplanade at least 66 ft wide was included on the seaward side of the allotments.

High water mark was much further beyond this.

Erosion of the beach has been progressive over the years, climaxing in the 1967 cyclone season.

Cyclones early in the year denuded the bathing beach. Unseasonable storms in June, and a long wet season extending from March–June, extended the erosion westwards to Noosa Woods.

In April Council carted 3000 cubic yards of sand to the beach in front of the bathing reserve. Sand movement on the beach is from east to west towards Noosa Inlet. Wave refraction round Noosa Headlands during rough S.E. weather contributes to the formation of a typical concave bay, Laguna Bay . . . .*

Where dunes have been faced with rock, erosion processes are accelerated and at the eastern end waves have virtually denuded the beach of all sand in front of Allotment 4 (Noosa Court), with deep water at the foot of the wall.

---

*The consulting engineers, in their wisdom, drew up a design in their Brisbane offices, ruling a nice straight line from the eastern corner to the Noosa Woods end. Even a child of eight knows that sea beaches have a natural curve, the apex roughly in the centre. The result of the straight wall is that sand manages to build up in the corner, where the rocks follow the curve of the beach, but is frustrated at the centre of the beach.

Above: *Long before the Hay's Island development (Noosa Sound). Looking along the Noosa estuary to Tewantin: Weyba Creek on left, Hay's Island in the middle distance and Mt Cooroy on the horizon.*

Below: *Another view of Hay's Island, showing the inlet behind the spit. This photo was taken from Picture Point in 1939.*

8

The rape of Noosa

I am afraid I now consider Hay's Island development to have been a tragic mistake.

Councillor Jack Hassett, then Deputy Chairman of the Noosa Shire

After the defeat of the high-rise building project at Noosa Heads, the Planned Progress Committee ceased to exist. The Noosa Parks Association was busy helping the Wildlife Preservation Society of Queensland to fight for the Cooloola National Park against sand-mining interests.

The small but vigorous local association was formed about 1960 when a big developer had persuaded the shire council that a road should be built round the Noosa headlands, so that 'old people and those unable to walk' could see the national park (and incidentally so that the company could develop some freehold land at Alexandria Bay). This was stopped, and now an increasing number of holiday-makers and many local people enjoy the easy walking tracks to the string of unspoilt beaches and the wild grandeur of Hell's Gates and Lion Rock.

While the battle of Cooloola was being successfully fought in and out of the Gympie Mining Warden's Court, and thousands upon thousands of postcards rained on the premier's desk asking him to 'Save Cooloola', the Noosa Inlet was lost without even a struggle. Dredging and filling, on a vast scale, altered the river's mouth irreversibly, creating problems that are only just becoming evident, and destroying the wildlife in that area.

In 1972, after the town plan was adopted, the council passed an application from Cambridge Credit (trading as Noosa Islands Estate, jointly with Intercapital Realty of Brisbane) to develop Hay's Island as a canal estate. Being partly inundated at high tide, the island was under the joint control of the Department of Harbours and Marine and the Lands Commission. Without a full environmental impact study and hydrographic study being carried out (this is denied by the developers, who brought in their own scientific team) the state government gave its blessing to the project.

The council, far from opposing the idea, was enthusiastic. Here was a 'useless' area of sand, mud and mangroves, a breeding place for biting midges, which was to be handed to the shire on a plate, all ready developed with roads, bridges, parking and building sites. It would be highly-rateable land with room for four hundred houses, connected to the mainland by two bridges, and all for the trifling sum to the council of the cost of half of one of the bridges.

At an estimated cost of $3 500 000 (a mere string of noughts to a large company with other projects in hand totalling more than $100 000 000) sand was to be pumped from the river to raise the overall level of the island by one metre. This was considered to be a safe margin above flood level. Of the fifty-nine hectares involved, less than four and a half hectares were to be kept for parks. The natural channels through the island were to be widened into waterways, using the river's flow to keep the channels free from stagnation and silting. Coincidentally, the council's sand pump, which had been declared too expensive to run, could now be leased to the development company.

Shire chairman Ian Macdonald was delighted with the prospect of all those new prestige homesites. 'The canal estate', he told the *Noosa News*, 'offers, at no cost to the ratepayer, addi-tional waterfront land of high value, which yields substantial amounts of rates and generally assists the economy of the area.' At a public meeting he declared, 'This is my baby', and added that the Noosa River had never had a serious flood 'in living memory'.

Yet quite a number of local residents were still around who had photographed the floods of 1931. Frank Bickle recalls when water was up to the steps of the old R.S.L. Hall in Tewantin. Howard Parkyn remembers the river 'racing across Hastings Street'. Mrs Olive Freeman says the view from Laguna Hill was of a sea of water at Noosa Heads as Hay's Island backwaters, meeting the run-off coming down the creek from the hill, overflowed into the site of the old paperbark lagoon.

In the 1893 floods there was not much that could be damaged at Noosa Heads, as the only residents were on high land; but, according to George Burgess, Lake Weyba rose nearly two and a half metres. In 1928, waves were breaking for the first time at Munna Point, halfway to Tewantin. In both 1931 and 1968, Noosa Heads was completely cut off, the approaches to both bridges being under water, and the river coming down at 10 knots.

However, there was much more room then for

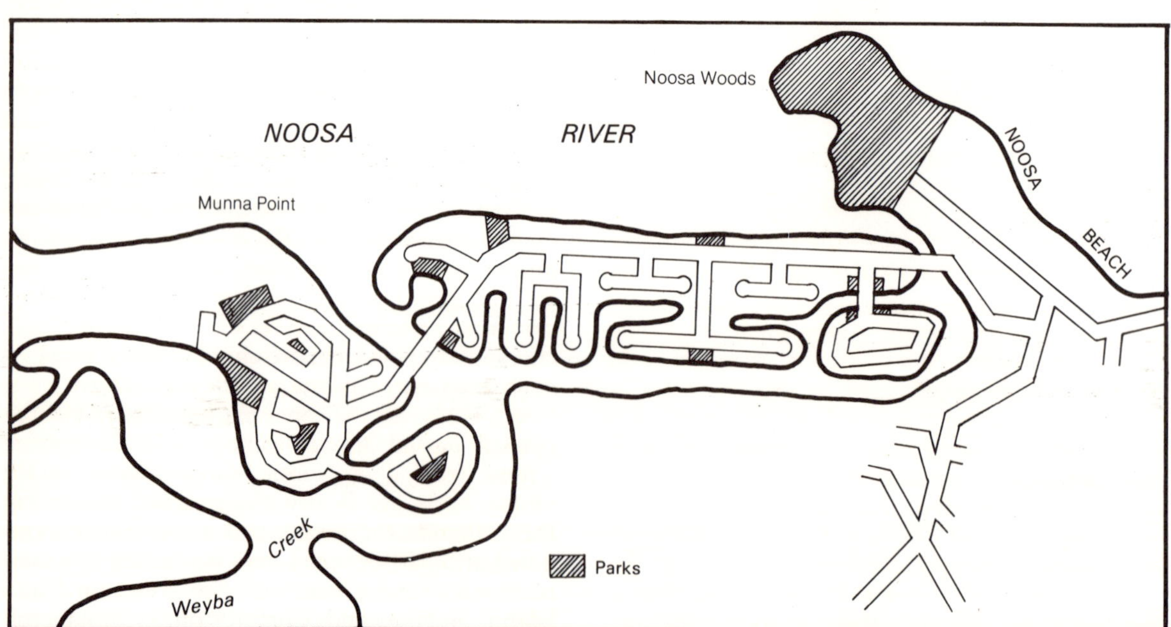

The Noosa Sound development. In 1979 work began on the connecting bridge across Weyba Creek to Portion 85, which formed part of Development Lease 8 and former Public Purposes Reserve 342. Reserve 342 was cancelled by order in council in 1956. Of the 70 acres, only 9 were kept as a reserve for recreation and camping. This is now the Munna Point Caravan Park.

Floodwaters from the river swirl across Hastings Street, at the Woods end, 1951. (By courtesy Denise Roberts.)

the river to spread out. Residents of Munna Point, who suffered flooding in 1976, believe that the problem has been made worse by the constriction of Weyba Creek by the sand bund put across the back channel of Noosa Sound.

The bund was constructed because the accelerated tidal flow through the dredged channels was undermining the first bridge, the only one connecting the new estate with the mainland. It was also scouring away at the point by Noosa Woods, where the council tipped yet more rocks to protect their camping area.

Many people were worried about the possible effects of a cyclonic surge, combined with high tides and a flood coming down the river, yet the developers knew that Noosa Heads was in a cyclone-prone area. Already the estate is becoming covered with expensive buildings.

Several isolated voices were raised in protest at the beginning, including those of local fishermen who feared the effect on fish stocks, and on the bar they have to cross each day. But fishermen have little political pull, and the more vocal conservationists were busy elsewhere. Now, wise

after the event, many are saying that Noosa Sound is a disaster which should never have been allowed to happen. Others felt, with justice, that they could not fight on *every* front.

Residents who have already built on the Sound will be liable for all future upkeep, dredging and restoration of canals, though they may not be aware of this. The shire council became alarmed by the growing costs of repairing the eroded bridge approaches, putting fill back on the front blocks after a cyclone, and shoring up the Point.

Parts of Hay's Island were always above most flood levels, as is shown by Commander Heath's plan of 'Nusa Harbour' made in 1870. He shows, among other things, fresh surface water at the Heads, rocks nearly two metres under the sand at the corner of the beach, and a stand of cypress pine (with '100-year-old trees' according to a contemporary timber man, W. Pettigrew) on Hay's Island.

This means that, when the land was cleared for the Noosa Sound development, trees at least two hundred years old were swept away overnight. Birds that had nested there for generations lost their homes, the white-breasted sea eagles moved into the tall gums at the bottom of the hill with the whistling kites, and thousands of distressed 'displaced persons' in the shape of mangrove honeyeaters, Lewin honeyeaters, azure kingfishers, mangrove warblers and silvereyes had to move up the hill, trying to find an ecological niche in an area already filled. (See Appendix A.)

For the developers did not clear by degrees. As soon as the connecting bridge was built, the big earth-moving machinery came in. The next morning, the residents on Laguna Hill woke to find a smoking ruin spread out in the estuary. It was as if a selective atom bomb had struck. Not one living thing was left, except for a large, solitary pine which must have escaped through some oversight. It soon died of exposure.

The amount of wildlife destroyed in this single operation is inestimable, if you count crabs, prawns, minnows, baby birds, lizards, snakes and insects. The whole life-support system of a complex, interweaving mass of bio-energy was swept away.

Perhaps the birds were luckiest; at least they could fly away, even if they had nowhere to go. Migratory birds coming back in the spring from Japan and Siberia were bewildered. They found a desiccated ruin instead of a green haven. Small dotterels and sandpipers ran about distracted, looking for a place to hide among the bare sand and 'instant grass', sprayed on complete with artificial green colour. The sea eagles and kites, both of which eat seafood, had to go far afield to feed their young.

Meanwhile, the Wildlife Preservation Society of Queensland and the Fisheries Branch of the Department of Primary Industries were mysteriously silent. New fish habitat reserves have now been declared up and down the river system. However, it is significant that these were delayed until after the Noosa Sound development was complete.

In spite of this, the Chief Inspector of Fisheries, Mr Harrison, is on record as saying that '. . . . Eighty per cent of coastal fish depend on shallow estuarine areas for breeding. The Noosa River system is one of the richest breeding areas in south-east Queensland. If these areas are dredged or disturbed, the number of fish would be greatly reduced.'*

---

*From the cutting books of Dr A. G. Harrold, secretary and founder of the Noosa Parks Association.

†Ibid.

Mr Les Ford, of the Department of Harbours and Marine, said that his department was concerned at the amount of development now taking place along the Noosa waterways, and that canal developments could prove a headache to future generations. Said Mr Ford:

We had not thought canal development would rear its ugly head in the Noosa River. Noosa Shire Council is in the position at the present time to avoid the pitfalls of this type of development, some of the problems of which are just becoming known. They need to commission engineers to do an overall report on the whole river. . . . full-scale, commercial dredging will not be encouraged in future, because the river could be subjected to tidal flooding and siltation; no one is quite sure.†

For the project to go ahead, agreement was required between the local authority, the Lands Commissioner, the Department of Harbours and Marine, and the Fisheries Department. In the light of Mr Ford's comments, it seems that some government departments were induced to compromise.

Said Council Chairman Macdonald:

The Fisheries Department and the Harbours and Marine have been aware of the Hay's Island scheme from its inception, and could have blackballed the whole project in its early stages had they wanted to. That they did not is, I think, sufficient evidence that the scheme by itself will not mean smaller bags of fish . . . as has been suggested.

Was it merely coincidence that in 1974, soon after Noosa Sound was opened by the premier, the government built a brand new, white-brick headquarters for the fisheries inspectors, in a prime position on the river bank at Munna Point? Silence is sometimes golden.

In December of 1973, the first stage of Noosa Sound was ready. Bitumen roads, duly kerbed and channelled, curved through the expanse of instant grass, now turning rather brown. A few imported Norfolk Island pines, umbrella trees and palms wilted in the strong winds blowing across the flat open space.

Premier Joh stepped from his helicopter like a rather undersized Father Christmas arriving for a children's party. Standing by the metal plaque on its stone slab, he announced that 'a former pest-infested area' was now a beautiful housing estate, and 'Cambridge Credit have nine projects in hand at a cost of $174 000 000. The company is showing its faith in Queensland.'

In an unconsciously funny speech, the Managing Director, Mr R. E. M. Hutcheson, announced:

Windsurfing has become a popular sport in the sheltered waters of the Noosa inlet behind the new beach of pumped sand.

The tailor are on! And it's standing room only at the mouth of the Noosa River near the new rock groyne.

(Photos: Lin Martin.)

Looking towards the North Shore from Tea-Tree Bay; calm waters framed by the spiky fronds of pandanus palms.

A pleasing pattern of waterways in the 1960s. Doonella Lake and bridge, right; Tewantin in the foreground. In the distance are Weyba Creek, Noosa Head and bar. Who needs canals? (Photo: Bernard Kuskopf.)

Pelicans at the Fish Board jetty, Noosaville.
(Photo: Bernard Kuskopf.)

The famed 'point-break' and Noosa beach. This picture was taken from First Point in the mid-1960s, before the rock wall was built. Only one guest house can be seen on the front. There's not a motel in sight. (Photo: Bernard Kuskopf.)

The rock wall has been built. It is low tide but sunbathers on Noosa beach are cramped for space. (Photo: Bernard Kuskopf.)

Hastings Street and the Noosa River bar from Laguna Lookout hill, early 1960s. Laguna House can be seen in the mid-foreground. (Photo: Bernard Kuskopf.)

Above: *In the mid 1960s Noosa Beach still preserved its natural curve. Boats shelter in the lagoon behind Noosa Woods. (Photo: Bernard Kuskopf.)*

Below: *The beginning of the end. The council's sand pump barge is altering the channel. The large sandspit was later swept away. (Photo: Bernard Kuskopf.)*

By 1968, indiscriminate piles of rocks have been dumped on the beach by private owners. The new Noosa Court Motel has been built out to high-tide mark. (Photo: Bernard Kuskopf.)

An engineer's dream of mathematical precision. By 1970 the wall has been straightened. Above it is an artificial esplanade of sand. (Photo: Bernard Kuskopf.)

Looking towards the North Shore, 1979, a view that owes more to engineers than to nature. The barren expanse of Noosa Sound spreads across the middle left.

The relocated river mouth, with its rock groyne, can be seen in front of the North Shore. (Photo: Jacky Mott.)

By May 1978, the appearance of the area has changed
dramatically. With every vestige of natural vegetation
removed, Noosa Sound development is reminiscent of a
wartime military installation. The new mouth of the
river, held in place by a rock groyne, appears to be
silting up, and a great wasteland of sand shields the
canal development. (Aerial photograph supplied by the
Surveyor-General, and reproduced by arrangement
with the Queensland government. Crown copyright
reserved.)

◀ Aerial view of Motel Hill and the Noosa Woods in 1967,
before the Noosa Sound development. Hays Island is a
mass of green vegetation. Note the deep channel near
the North Shore. (Photo: Bernard Kuskopf.)

The appeal of Noosa has many facets —
No cars allowed: walkers enjoy the quiet and safety of
the track through the national park. (Photo: Jacky Mott.)

The soft, creamy blossom of a flowering eucalypt.
(Photo: Jacky Mott.)

The trunk of a scribbly gum appears to have been etched
by some eccentric calligrapher. The marks are, in fact,
caused by a burrowing insect, Osmographtis scribula.
(Photo: Lin Martin.)

Pebbly foreshores, Noosa National Park; looking towards the North Shore from Granite Bay. (Photo: Lin Martin.)

Halse Lodge, formerly Hillcrest guest house, preserves some of the charm of Noosa's earlier days. It is now hidden behind the four-storey Breeze development. (Photo: Jacky Mott.)

Once an area of wildflowers and wallum, Noosa Junction is a desert of tar and cement with no aesthetic charms.

The Bank of New South Wales building in Hastings Street. The last blue gum in the street was cut down to make way for this concrete canopy.

Must an electrical installation always stick out like a sore thumb?

(Photos: Jacky Mott.)

The Noosa Heads Hotel Motel. The original low building, designed to hug the hillside, has been dwarfed by this brick monstrosity. (Photo: Jacky Mott.)

A welcome departure from concrete and brick: the Attic Restaurant with its wooden gables and stonework. Since this photo was taken, new unit developments have blocked the restaurant's view of the sea. (Photo: Jacky Mott.)

Completely lacking in imagination or aesthetic quality: the new Villa Noosa Hotel-Motel at Noosaville. (Photo: Mike Noone.)

'Instant palms' go up at Ocean Breeze units, with the help of a crane. Behind these four-storey buildings is Halse Lodge and its green lawns (see colour plate, earlier page).

Little Cove (Johnston's Bay) from First Point, looking towards the North Shore. (Photo: Bernard Kuskopf.)

High-density unit development, Noosa Sound. Stark lines and red roofs make no attempt to blend with the riverscape.

(Photos: Mike Noone.)

Noosa beach in 1981. The green backdrop, which is freehold land, is to be acquired by the Noosa Shire Council to prevent building on this fragile and very steep slope. (Photo: Mike Noone.)

The 'new' Noosa beach in 1979. Waves pound away, eating further into the three-metre cliff of sand. (Photo: Jacky Mott.)

My company believes in conservation . . . . We don't apologize for taking this land away from the mosquitos and the sandflies. *We have turned an ecological desert into a fine estate* [author's italics].

Then, he and his eighty guests sat down to a gourmet luncheon, after which they were whisked up in helicopters to admire the neat loops of canal and lawn.

That the company's faith was misplaced, not to mention the faith of shareholders in the company, was soon revealed when the whole house of cards collapsed in the following year. By August 1974, Cambridge Credit and Intercapital Realty were in the receiver's hands, and thousands of investors had lost their money. Work stopped overnight. Dry sand that had been pumped for the second stage began to blow back into the river. The vital third bridge could not be built unless the shire council footed the whole bill, so the estate was left a dead end, with subsequent traffic congestion in Hastings Street. (This third bridge has now been built.)

With the sale of some of the first developed area, finance was found to continue the second half of the project. The lovely little crescent of sand south of where Weyba Creek enters the estuary was swept away and dredged out, while the green space behind Munna Point, opposite, was once more bulldozed to bareness. This became an expanse of sand fenced off with concentration-camp thoroughness by high barbed wire. On all early maps of the district, the whole 70 acres (28 hectares) of Munna Point is shown as a public purposes reserve (R. 342).

Within months of commencement of pumping operations, according to Kathleen McArthur,* layers of black, detritus-filled sand and silt had been carried by the tides to cover the white sandy beaches on the south-east corner of Lake Weyba. She deplores all such dredge-and-fill operations, which break the seal that forms under any body of water, allowing the disturbed detritus to move back and forth with the tides. 'By the time the whole planned scheme has been constructed, the river between Noosa Heads and Munna Point [will have] been ecologically destroyed. It will then be too late for redress.'

At the same time as the Hay's Island development was getting into gear, Coastal Rutile was busy destroying the foreshore vegetation on the

North Shore. Salt winds swept in over the flattened and denuded dunes, killing the 'third line of defence', the paperbarks in the small freshwater swamps between the beach and the Noosa River. This company shortly afterwards lost its deposit for not rehabilitating the dunes properly, and its leases have now expired.

At Hay's Island, a fringe of mangroves had protected the higher land, preventing erosion by slowing the tidal flow with the filtering action of the roots and trunks, and allowing sand in suspension to settle. Sand-pumping has the opposite effect. It stirs up the silt and sand, deepens the channels, speeds up the tidal flow, and allows ancient detritus to drift over valuable seagrass beds and smother them.

Following the disaster to the birds came the disaster to the warm, shallow fish-breeding grounds, the home of small whiting and flathead. Mangroves are now known (and were known then) to be a rich source of nutrients. About 92 per cent of fish and 99 per cent of prawns (by weight) that are caught commercially are dependent on mangroves for some part of their life cycle. Bacteria break down leaf debris, which makes food for small aquatic animals such as bloodworms, snails, limpets and crabs. These provide food for juvenile fish when the tide is in, and for large wading birds when the tide is out. Fingerling fish, in turn, provide food for eagles, ospreys, pelicans and larger fish.

Already in the Noosa river system, 75 hectares of wetlands have been appropriated for other land use purposes, and another 135 hectares are earmarked for draining or filling. This leads to increasing velocities of tidal flow and causes bank and bed scour, according to the government's Coastal Management Inquiry.

However, all seemed well until January 1974, when Cyclone Wanda came down the coast, bringing disastrous floods to Brisbane. At high tide, the water in the new canals was just lipping the edge of the lawns, and the new bridge to the Sound began to erode at its base. Noosa Woods camping area began to wash away. Rocks and clay were hastily tipped on the sea-facing side of the main island, and on the end of the Woods area.

The question then arose: Who was to pay for repairs when storms or floods damaged the bridges and foreshores of the new estate? The developers had handed it over to the shire, which was now left holding the baby.

Then, in January 1976, Cyclone David battered the whole coastline. At high tide, waves were

---

*Kathleen McArthur, *A Living River — The Noosa* (Caloundra Branch, Wildlife Preservation Society of Queensland, 1974).

breaking over the top of the rock wall at Noosa Heads and swirling into Tingirana Arcade. It was feared that the sea might break into Hastings Street itself.

On the island estate, waves broke over the unprotected front building-blocks with their flimsy rock edging, washing away soil and killing the lawns with salt. The bridge was once more in danger. Water flowed across the road at the old post office. A sand bund had been put in near the mouth of Weyba Creek, to stop the river scouring through the back channel behind Hay's Island; but Munna Point residents complained that it then backed up the creek and flooded their homes. The council demanded that the front blocks at Noosa Sound — those most vulnerable to the sea — be withdrawn from sale.

In April the premier was invited back to have a look at the disaster area he had launched with his blessing only two and a half years before. After promising some help, he then walked 'the length of Noosa Beach', which at high tide meant about thirty metres.

The shire, through its sitting member for Cooroora, Mr Gordon Simpson, had now thrown the baby back into the government's lap. The Minister for Tourism and Marine, Mr Hodges, made an inspection also, and said that the valuable investments on Noosa Sound must be protected at all costs. Mr Simpson asked for urgent action before the next cyclone season; but the government was busy preparing for an election.

It was known that the Department of Harbours and Marine had called for a detailed hydrographic study of the river and bar, already carried out by Professor Gordon McKay of Queensland University, and that he had made several recommendations for rehabilitating the beach and protecting the canal estate, including sand-pumping, and possible retraining of the river mouth.

This plan was adopted in December of 1977 by council, after long haggling with the Lands Department over who should foot the bill. It was finally agreed that the state government would find $400 000, the developers of the Sound $400 000 (from expected sales of waterfront land) and the shire the same amount.

As the plan included a rock groyne and straightening and deepening of the river mouth, which would be sited further to the north-west, local fishermen became uneasy about the possible effect on the river and bar. They wanted a temporary, easily removeable groyne of sand.

However, the resiting and training of the river, so as to protect the Noosa Sound development from the open sea and the river currents, has now been completed at a cost of $1 400 000, and the fishermen's worst fears seem to have been realized. The river appears to be silting up and boats have trouble getting into the upper lakes.

There is an irony to it all. If the council had spent only $100 000 on pumping sand over the beach in 1970 when it first bought the pump (instead of allowing the Hay's Island development and letting the sand be pumped to raise the island) the beach could have been restored for one-quarter of the cost, and we would not be having problems with the river flow. It was a case where a stitch in time would have saved nine — and would have saved the beautiful Noosa Inlet from rapine and destruction.

The earliest-developed blocks on Noosa Sound are now covered with individual houses and gardens with new trees that have improved the former bare, windswept look. But at the Munna Point end, an eruption of multiple dwellings in the form of strata-title units has filled the flat reclaimed land, still treeless and bare, and bordered by a fringe of protective rocks.

The Noosa estuary is now like a mouth filled with too many artificial teeth. Many of the buildings are reminiscent of Alcatraz or Pentridge Gaol, and many of them appear to be empty. However, they are still changing hands at a great rate for 'capital gain' (the latest fever in the Noosa area, which is beginning to look like a gold-rush or the maddest days of the Poseidon mining boom). Seafront blocks in Hastings Street, although there is no beach left in front of them, are bringing well over a million dollars. The luxury First Point units are reputedly let for $1000 a week — in the off-season. Land prices have risen steadily by an average of 25 per cent per annum.

Real-estate writers in the southern newspapers have been critical — perhaps jealous? — of this continued boom. In 1980–81 they began writing gloomy articles forecasting the possible bursting of Noosa's bubble, with many investors being caught. But so far there is no sign of the 'bust'. Speculation in town planning consent is rife. Landowners apply for special consent to build a multiple dwelling, and as soon as consent goes through they sell again at a greatly enhanced priced. Investors have made $20 000 to $30 000 overnight in this way. Even when the units are built, they immediately change hands again to someone who wants an income-producing investment. Most are owned by absentee owners.

9 Development disasters: the uglification of Noosa

Stirred by an uneasy feeling that Noosa Heads was no longer the same beautiful place that had first attracted them to settle there, some local residents recently formed a Noosa Beautification Committee. Alas, their efforts are like those of a cosmetician applying rouge and powder to a scarred, once-lovely face. A few little plots of zinnias, some pot-plants and rock gardens, cannot hide the basic ugliness of places like the Noosa Junction shopping area.

This is the introduction to Noosa Heads for visitors coming from the south by the Noosa Coastal Highway: an increasing number of yellow-brick buildings blotting out the view of the sea to the right; a scatter of advertising signs on the left, becoming larger and more strident, extolling the virtues of the caravan parks and motels; then the beginning of ribbon development leading to the junction with the Noosa–Tewantin road.

Noosa Junction, from a single application for a hardware store in 1969 (rejected at that time as the area was not planned to be commercial) has grown into a large shopping complex with banks, chemists, bakers, general stores, restaurants and a delicatessen. It has recently spread round the corner of Cooyar Street into what was an area of wildflowers and wallum at the back of Cooloola Hill.

The place has grown in haphazard fashion. Parking arrangements are chaotic, and the Commonwealth government added to the congestion by resiting the busy Noosa Heads Post Office there. In 'rationalizing' its services, it has closed both the Sunshine Beach Post Office and the one at the foot of Motel Hill, causing residents of both beach resorts to make a long inland and uphill journey to post their mail.

Council demands for off-street parking have led to a desert of tar and cement. Electric light poles,

Hastings Street, Noosa Heads circa 1947. (By courtesy B. Lintott.)

concrete kerbing, traffic islands, garish signs, petrol stations and car yards, complete with plastic bunting, add up to a typical scene of Australian ugliness.

At Noosa Junction most of the trees are gone. And this on a corner where seepage from the high sandhills above, and the protection given by those hills from salt winds, had led to an area of dense greenery, lit with wildflowers each spring. The old-man banksias, the intense green of Cooloola pines, the feathery fronds of lions'-tails and casuarinas, the green spears of blackboys (grass-trees), the scribbly-gums and bloodwoods covered in blossom in summer, have been swept away as though they never existed, and replaced by ugly electricity poles.

No proper footpath was left on the north side, and cars turn and travel in all directions through what should be pedestrian areas. The siting of the huge Coles Supermarket and Noosa Fair shopping complex right alongside the Junction has increased the problems to a point where something must be done. Australia Post moving round the corner to the Noosa–Cooroy road has alleviated some of the parking density, but has led to cars making

dangerous U-turns on the highway.

The council's engineer and town planner have drawn up several plans for improving the junction, including one for a new driveway between Pinnaroo Park and the Bowls Club to take some of the congestion from the main road.

The town planner admitted that it was a compromise solution, the one that would cost least to implement — short of doing nothing. The extension of Lanyana Way (the most obvious solution) would involve buying back some expensive land from the Uniting Church, which is doing a deal with a big developer for the use of this site — bought for a few hundred dollars many years ago for a church building.

The Banksias Caravan Park on the corner has already been bought for a grandiose scheme involving a six-storey building and underground parking for 1000 cars. Another huge development planned is a $3 million plaza of four storeys (really five) on the corner of Lanyana Way — a dead-end street. As usual, the architects' fanciful drawings show giant coconut palms towering to the roof level.

Of course, there are some good things. The

First Point: the trees fall to a developer's bulldozer. (By courtesy Noosa News.*)*

Casa building with its Spanish arches, designed and built by Brian Coutts to house a restaurant (now closed) as well as shops, has an adjoining courtyard filled with old banksias full of character, while a few graceful paperbarks shade the other end. The cobbled courtyard across the street, once the garden of a small motel, has become an attractive open-air setting for a coffee shop and a cluster of art, craft and music shops.

Two arcades lead through from the main street to Lanyana Way and another major disaster: a vast, bare car park for the Noosa Fair shopping complex. This complex could just as well have been sited in some of the cleared, bare land near Tewantin; but the developers wanted the magic name of 'Noosa', and the council wanted their $900 000. Since the site is only one block from the main entrance road to Noosa Heads, it was known that traffic would increase — and at present the car park is a dead end with only one entrance, that from Lanyana Way.

Macbeth's witches propounded the paradox 'Fair is foul, and foul is fair', which might be applied to the Noosa Fair shopping town: an attractive enough concept, with its artificial lakes and palms, set in the wrong place. In spite of bitter opposition from residents, and from more conservation-minded councillors, it was built on the edge of one of Noosa Heads' most pleasant residential areas, and is an eyesore from the prestigious high blocks behind the Noosa Heads Hotel. Half the hillside was dug away, the sand carted off and sold, to make room for a flat expanse of bitumen parking and the red-brick box of Coles New World Supermarket. Three hectares of dense bushland, big eucalypts and flowering banksias were devastated overnight by the giant bulldozers, which can tear the trees bodily out of the soft sand. A shocked possum with its tail almost severed was found next morning on the highway and taken to the vet for treatment. A bewildered koala was found sitting in the middle of the old Weyba bridge. These were two that survived — hundreds must have perished. The birds lost their nests and their young.

Some of the opposition came from local retailers, already struggling with high rents, seasonal fluctuations and a marked fall-off in tourist numbers in 1979–80. The local Chamber of Commerce mounted a massive and expensive campaign to 'sell' Noosa in the south — something that had never been needed before (good wine needs no bush), and by Christmas 1981 the tourists were returning, but to an altered Noosa Heads.

Motel Hill has been caught up in the current explosion of 'multi-unit' development, and in place of the modest cabins set among native trees at the Noosa Gardens Motel there are now multi-storey buildings, while Edgar Bennett Avenue and Viewland Drive residents lose their view.

Then comes the disastrous design of the red-brick Noosa Hotel-Motel on its magnificent site overlooking the sweep of Laguna Bay and the North Shore; but no longer the highest building on the hill, where it used to stick out like a sore thumb.

On the left, the attractive Attic Restaurant, with orange-flowering creepers covering its wooden gables, has been cut off from the sea by a tasteless jumble of concrete towers. The run-off from all these buildings could cause some interesting developments when the monsoonal rains cascade into Noosa Sound below. Already a whole new creek has started to form near Coral Tree Avenue, through the run-off from Nairana Rest and the Quarterdeck — steep hills now becoming fully built over.

The area known as Picture Point, above the hotel, was declared a sand-dune problem area in 1977. This means that no one shall erect a building, or clear the land of natural flora, without a written permit. Several such problem areas have been declared, one of them the steep area above the beach at the beginning of Park Road. The council is now applying to rezone this land 'open space' so that it can never be built on, enabling Noosa Heads Beach to keep its green

The white gate is the old entrance to Colonel Lintott's hilltop property in Viewland Drive, overlooking the Noosa River bar. Another sort of bar is here now: the Noosa Hotel and its drive-in bottle section would be just to the left in this picture. (By courtesy B. Lintott.)

Above: *Excavating in pure sand: constructing the road round to the Noosa National Park, 1947–48. (By courtesy Kevin Freeman.)*

Below: *The opening of the Noosa National Park, 1949. Sir John Lavarack performs the ceremony. (By courtesy Robinson Studios.)*

Sign confronting residents attempting to walk to the north end of Little Cove beach from the road. The sign was well outside the First Point development's boundaries. (Photo: Dennis Massoud.)

backdrop. They had been warned by what happened in Natasha Avenue, a section of sandy hillside above Noosa Sound, where a whole section of hill collapsed. Residents in a block of units woke to a roar of falling earth, and found themselves suspended over a four-metre void. Needless to say, this is now zoned a 'problem area'.

Mendoza's long, low building on Motel Hill is still unchanged, a piece of the older, more tranquil Noosa with its bougainvillea-draped verandas. Further down on the right, the low buildings of Ocean Breeze Motel have been replaced by a development disaster that has left old-time residents stunned. The whole block, including the lawns and hibiscus gardens, has been built over; to the very borders of the road it is filled with brick, block, stone and cement. Picturesque old Halse Lodge (soon to be developed by the Church of England) and the famous gum-trees where the eagles nest are hidden by four-storey towers with red-tiled roofs, like scarlet-topped toadstools.

The pile of rocks and earth about the base, placed there at enormous expense, is to cover the ground-floor garages and comply with the town

Another man-made disaster: four-storey buildings of Ocean Breeze units built right to the edge of the footpath, almost obscuring Halse Lodge and its famous landmark gum-trees. Compare this with the colour plate of Halse Lodge. (Photo: Mike Noone.)

plan consent for 'three storeys over underground parking'. This is the latest ruse for getting away with another storey: you build garage space above-ground and pile earth around it; it then becomes 'underground' parking. How the present site cover and height were allowed by the council's building department is a mystery and a tragedy. A rash of 'instant palms' has gone up in front of the rocks and bare walls. There is a positive epidemic of these spindly palms in Noosa Heads. They are not even Australian palms such as grow in the local bush, but imports from South America. Nurseries grow them because they have a small root system and quite large palms can be transplanted successfully. As far as shade goes, they are about as effective as a stick of celery.

Across the road, behind the little bit of space (council-owned) that remains for car parking, another shade tree which was part of Noosa Heads' skyline for many years has fallen to the developer. This huge Moreton Bay fig (with a trunk five metres in circumference), last refuge of a colony of fig-birds, has been cut down by Kern Bros to make room for sixty-eight more units,

more shops, more restaurants. . . . The land, which was owned by the controversial Noosa Resort Corporation, changed hands for $4 million.

Angry reaction in Division Four followed the council's decision to grant approval for four town house units with separate titles to be built at First Point, the little rocky headland to the right of Noosa Heads Beach, before Little Cove. The cry of 'Save First Point!' was heard, for this was the first intimation that any land on the seaward side of the national park road was privately owned. It had always been regarded as public land.

The owner, who lived in Sydney, then offered to sell her block to the council for $100 000. The council said it had no money, but would produce $40 000 if the residents of the shire could put up $60 000. An appeal was launched, but it was quite unrealistic to expect a small community to raise such a sum by private subscription, though there were several guarantees of smaller amounts. Meanwhile the asking price went up to $120 000. The owner asked for alienation of another strip for parking and entrance, but this was refused as

Land slip at Natasha Avenue, Noosa Heads, on March 20, 1977. Ayer Tara, a block of four units, is suspended over the void above Noosa Sound. Soil experts said a whole section of hillside had apparently slipped. (This area was not included in the 'sand-dune problem areas' gazetted by council in Division Four of Noosa Shire.) (Courier Mail photo.)

the space would be needed later for widening the national park road. Finally, the owner sold to a developer. The land, overlooking lovely, 'unspoiled' Johnston's Bay or Little Cove, has been torn apart, its cover of paperbarks, gums and bloodwoods felled to make room for a building which covers almost the whole block.

The price of the land had eventually gone up to $150 000. Before approval for multi-units was stupidly given, the land could have been acquired by council for less than half the first asking price. A huge stormwater drain at the back of the beach now discharges run-off from the built-over Laguna Hill above. Little Cove can no longer be called 'unspoilt'; and First Point is lost forever.

Just over ten years ago, council could have bought the open space which remained for so many years opposite Laguna House (site of the Laguna Shopping Arcade now) for £6000. This used to be an alternative access to the beach. It is now filled with units that are selling for more than $100 000 each, and there are thirty units.

Hay's Corner, opposite the beachfront car park, which could have been kept as a tribute to Noosa's first pioneer (since Hay's Island has now become 'Noosa Sound') has had its old Queensland wooden house on stumps removed, the site filled with many thousands of tonnes of sand (which might better have been put on the beach), to take an art gallery and more shops. The old mango tree has been kept and gives its name to the pleasant outdoor cafe on the corner.

Tewantin, once the centre of business and tourism, suddenly found that the small village of Noosa was a tail that had begun to wag the dog. In the late 1960s and the '70s many branches of Tewantin businesses were opened at Noosa Heads, though Sunshine Beach remained a small satellite town.

The Bank of New South Wales lost several customers when it decided to replace its Noosa Heads branch with an impressive new building further along Hastings Street. The block contained an old house and a great many trees, including the last remaining blue gum in the main street, which provided shade for cars and pedestrians.

In spite of protests to the (temporary) manager and urgent appeals to Brisbane head office, the gum tree, the casuarinas, and everything else except a fig tree, came down to make room for an egregious embellishment designed by the architects Cooke and Kerrison: a walkway with a huge overhanging canopy of concrete where the tree had been. The building itself was set six metres further back.

Opposite the store and the first post office was another, larger, fig tree. This was poisoned by someone in 1969. Tingirana Arcade was built *around* a gum tree, an old Moreton Bay ash, now embellished with a green elkhorn fern.

In 1970 the council decided to clean up the car-parking area above the beach with bitumen lanes for cars. Council bulldozers were about to remove every living thing (they ended up taking the banksias and cupanias, glossy-leafed natives which give a dense shade) but a few enraged residents confronted the bulldozers and demanded to see the shire clerk. The big bloodwoods by the shelter shed at the back of the parking area were saved.

A few pathetic slash pines were then planted on the beachfront, and rapidly died. Two casuarinas have now established themselves, living through cyclones Wanda and David; and, of a row of new

"FLAMIN' foreigners — they're taking over the country."

Cartoon by Jeff, Sun News Pictorial, *Melbourne, August 19, 1978. (By courtesy the Herald and Weekly Times Ltd.)*

cupanias, all were flourishing until six were stolen by a mean-spirited visitor over Easter 1977. On the advice of native-plant expert Kathleen McArthur of Caloundra, council workmen planted hardy coast banksias, exactly like the ones they had cut down, and these have survived.

There are no effective provisions in council by-laws for preserving even the largest and oldest trees, unless in a 'sand-dune problem area'. A

116

token gesture towards preservation is made by attaching a printed slip to each building permit, urging owners to 'preserve native trees where possible'. This naturally is ignored by most people, who find it more convenient to get a man with a bulldozer to clear everything while levelling the block. Even when the owners do wish to save some shrubs or trees, bulldozer drivers, who are notoriously destructive (as inevitably so as a child with a large pair of scissors), will often knock them down 'by accident'.

Clear-felling is still going on, even on such steep hillsides as those of Nairana Rest and Eugarie Street. The fact that every block has not yet been built ·on gives an illusory wooded look to this area, but in ten years it has changed from a forest of native trees to a forest of electricity poles, transformers, and television aerials.

The shy green-winged pigeon, the booming pheasant-coucal and other ground birds have moved out or been killed by domestic cats. Cheeky magpies, kookaburras and pied butcher birds have moved into the new open spaces, becoming expert beggars as they perform a few aerobatics in exchange for a regular hand-out of cheese or meat.

Most blocks are so narrow (the average area is 20 perches, with twelve to fifteen metre frontage) that once a house has been built it fills the whole width, as owners are obliged to leave only one and a half metres to their borders by council regulations. This means there is no room between houses for any of the original bush. (See Appendix B.)

Much of the development in Noosa has been to the good: there are better roads and accommodation (or at least more of it), and a variety of excellent restaurants serving dishes from Spain, France, Italy, the Middle East. It is no longer necessary to go thirty-eight kilometres to Nambour or sixty-four to Gympie for household

Mrs Bowden's 'Dialogues' group, circa 1929. Frank Bickle is seated, second from left. (By courtesy Frank Bickle.)

goods, carpets, furniture, fashionable clothing and car parts.

Noosa Heads is, in fact, a cosmopolitan place considering its small population. A Melbourne elite has chosen to make this its annual wintering place, and many of those who have made a yearly pilgrimage to the warmth and clear air of Noosa come here to live when they retire. Many expatriates from warmer climates — New Guinea, Malaysia, Africa, Burma, India — have settled here. So many locals are migrants from Adelaide, Sydney, or Melbourne that they complain that they 'never manage to meet a Queenslander'. The first person a fugitive from Melbourne's winter is likely to meet in the main street is someone from his bowling club back home in Beaumaris or Camberwell. As the sun comes out, so do the bowlers in their white gear, like some exotic form of bird which appears only when the sun shines. There is also golf, with an eighteen-hole golf course at Tewantin, and tennis, with electric light courts.

The Noosa River Sailing Club has been in existence for twenty-five years. At first the members used to meet under the shady cotton trees on the river bank at Noosaville. Then the long, roomy clubhouse was built, which is used for all sorts of community activities in the off-season from September to May. There are now thirty-odd boats, and fifty-one active members. On Sundays of brisk wind, the blue waters of the estuary are alight with coloured sails and ballooning spinnakers.

All this, with the leavening of easy-going Queenslanders who still live at Noosa, has led to a lively but relaxed social round. There is a good shire library, a state school and a kindergarten at Tewantin, and the Noosa District High School at Cooroy.

The Noosa Sound brochure, a coloured production on glossy paper, begins:

Just ninety miles from Brisbane lies one of the most beautiful and natural beaches in Australia . . . . There's year-round surfing with temperatures about 20 to 30 degrees . . . . far enough away to escape the highly commercialized beach resorts of the south, yet sophisticated enough to provide good living, including excellent restaurants.

True enough up to a point, if you include the beach south of the headland. Noosa Heads beach could hardly be described as 'natural' any longer; and the temperature almost at the water's edge at Peregian can drop to freezing point on the grass in mid-winter. For, like Brisbane, this is still well south of the Tropic of Capricorn.

The Noosa district attracts artists like a magnet: from Boreen Point to Eumundi to Peregian and everywhere in between, painters are drawing inspiration from the blue days and starlit nights of one of the best winter climates in the world. Besides the professionals, there are innumerable amateurs daubing and potting and modelling and thoroughly enjoying themselves.

With only a drive-in cinema, and before that a small wooden picture theatre at Tewantin for entertainment, locals have always dabbled in amateur theatricals. These have ranged from Mrs Bowden's 'Dialogues' group of the 1920s in Tewantin and Gwenda Marriott-Burton's drama group at Noosa Heads in the '40s, to the flourishing Noosa Arts Drama Association (NADA) of today, with its own small theatre and meeting hall near Weyba Creek Bridge. 'Noosa Arts' is as much a social club as an arts group, though it holds an occasional art show. The real artists tend to be loners. The late Max Newton's remaining pictures were donated to Noosa Arts by his brother Bert Newton, of Melbourne, and a memorial plaque to Max is to be put in the foyer of the new extended theatre, now being built at a cost of $58 000. Peregian, too, has a flourishing theatre club and pottery club.

Division Four of the Noosa Shire, embracing Tewantin, Noosaville, Noosa Heads, Sunshine Beach and Peregian, has grown phenomenally in recent years. In 1910 when the shire was formed, Division Four did not exist, and the total rateable value of Divisions One, Two, and Three was £169 380. Today the value of council income is over $2 000 000, and several million dollars' worth of new buildings are approved each year in the shire. The growth rate for both population and expenditure is better than the Gold Coast and is far ahead of the Australian, or even the Queensland, average.

This rapid growth has resulted in a strain on local services: roads, sewerage and water supply. Some retired people, whose lives are now centred round their gardens, have been chagrined by savage water restrictions causing valuable lawns and shrubs to be left to die unless laboriously watered by hand. This in an area of high summer rainfall, and a full dam — simply because the reticulation is not adequate.

A plan for augmentation of the water supply has now been drawn up, at a cost of just under $8 million and spread over ten years. Water storage will have been increased five-fold when the operation, including the raising of Six-Mile Dam, is complete. Better reticulation and more storage

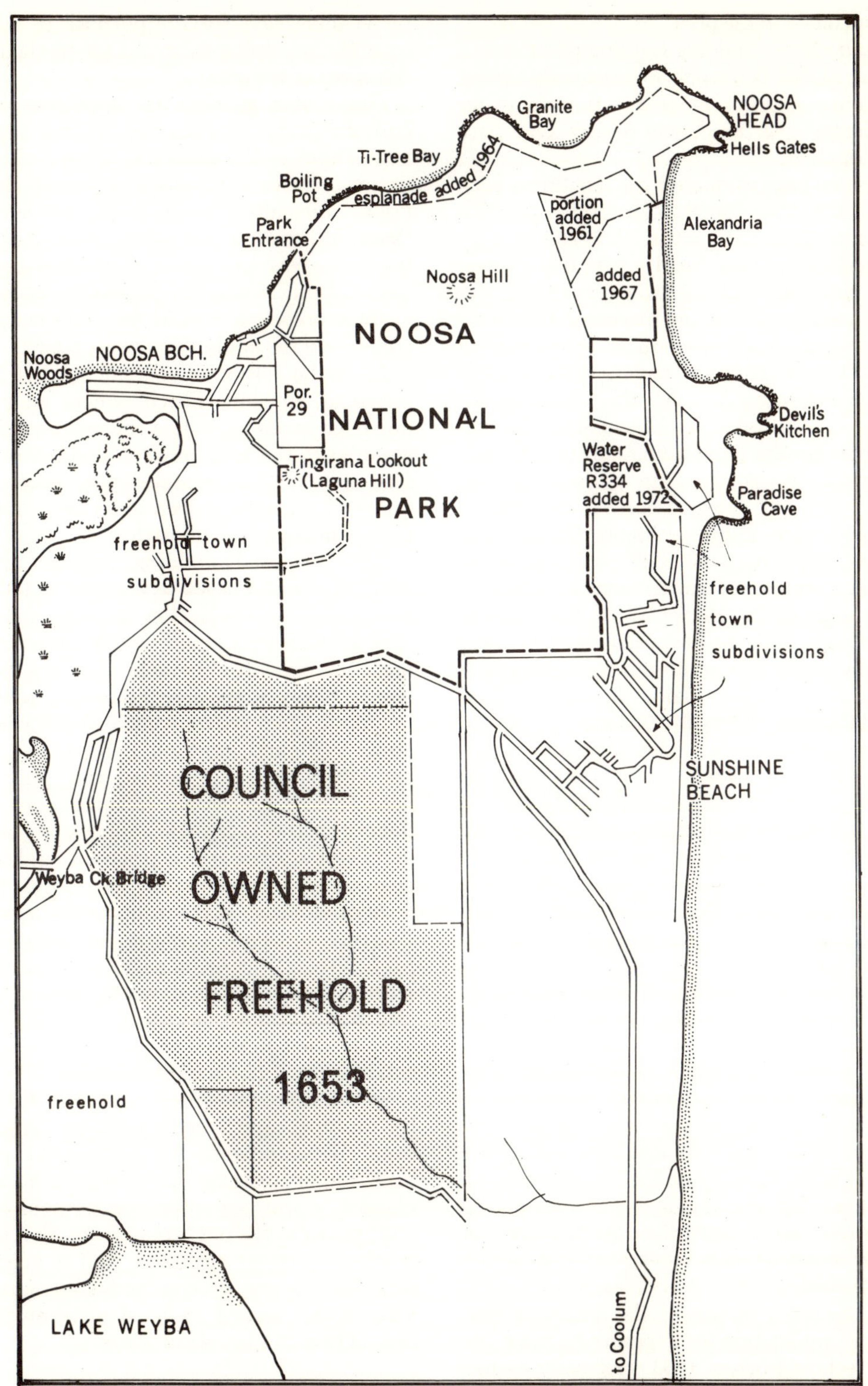

Council-owned freehold Portion 1653, which the C.M.I. report recommended strongly should be included in Noosa National Park. An application to mine peat there was rejected in 1978. The access road to the new Sunshine Beach school now runs through it, and the council is planning limited housing development nearby. (Drawn by Dennis Gittoes.)

tanks are included in the plan.

At the same time, building is going on at such a rate — over $1 000 000 worth of approvals having been given in one month in 1978 — that the water supply, already overtaxed, will hardly keep pace with development.

Water costs have escalated by ten times each year from 1973: from $3008 to $36 600 in 1974–75 and from $36 600 to $337 180 in 1975–76.

Discontent came to a head when the Valuer-General's new land valuations were released in 1978, in some places showing increased values up to 1500 per cent, with consequent steep rises in rates. The Division Four Ratepayers and Residents Association determined to ask the Governor-in-Council for dissolution of the Noosa Shire Council and the appointment of an administrator. To this end a petition was circulated among residents and presented through the Local Government Minister, Mr Hinze. Public feeling is running high in Division Four, the most highly-rated and the most vocal part of the shire, over all sorts of issues: the state of the Noosa River, the possible cost of Noosa Sound, the proposed Noosa Waters canal development, the resumption of properties for raising the Six-Mile Dam, high rates, First Point, and the costly new shire offices.

The trouble with many council policies is that they lack continuity. When the town plan was introduced way back in 1972, it contained a section relating to problem dune areas, yet nearly five years passed before any move was made to declare the obvious problem areas in the shire. The whole of Sunshine Beach was declared a beach erosion control district in 1972 by the Beach Protection Authority.

The functions and powers of local authorities are far greater in Queensland than in other states of Australia. Town planning, land subdivision, provision of essential services, and the control of coastal erosion zones all come under the mantle of shire councils, besides recreational and cultural facilities, rubbish removal, and noxious waste disposal. Thus the 'quality of life' in a developing area is largely in the hands of a few elected representatives, not necessarily experts in any of these fields, and often with private interests which may conflict with those of the community.

One of the latest threats to the local environment was an application for a 'permit to enter' the council's freehold Portion 1653, originally owned by Walter Hay, for the purpose of mining peat from the low-lying swamp, a former lake bed, that fills most of the area.

Had it been granted, the applicant would have had the right to excavate all the top-soil and vegetation, as the swamp is all peat. It is also the drainage basin of Burgess Creek, and absorbs and holds rain after heavy falls, preventing flash floods alternating with a dry creek. As it is one of the few areas left where Christmas bells can be found about Noosa, the Noosa Parks Association at once objected. So did the shire council, to its credit, and the application was refused.

A similar situation to that at First Point had arisen at Sunshine Beach, where an application had been made to build a four-storey, twelve-unit complex in the area between Webb Street and the sea, right on the frontal dunes. There were two hundred separate objections from residents, and the application was refused.

The proposed building negates all modern thinking about seaside development and, to make matters worse, the land in question was originally earmarked for parkland, and is shown as such on the 1928 plans for the area. The one hectare between Webb Street and the beach, which should be public open space, somehow became freehold about 1970, and was built on by a Mr R. Petersen, an American, who exulted that such absolute beachfront land 'would have cost me a million dollars in the U.S.'. Of course such land in the middle of a town area is priceless, and should not be privately owned.

Another area of four hectares between Ross Crescent and Park Crescent was also to be park. Mr E. Webb, former manager of T. M. Burke Ltd, when asked whether it would remain parkland, said that it would, and that it would be handed over to the council. This is now all alienated for subdivision, if not actually built on.

So the sorry tale goes on: permits for high rise, for multiple units, for building on foredunes, for canal developments; no policy on preserving bushland or trees; no water for keeping the exotics planted in the new suburban gardens alive; fish and prawns all but gone from the lakes and beaches; boating curtailed, and a huge debt of $400 000 to the Queensland state government. This is the record of local government in the destruction of a beautiful place.

10 The Coastal Management Investigation

The Noosa River and its associated lakes probably constitutes the most important river and lake system, from the aspect of recreation and conservation, in Queensland.

Coastal Management Investigation Report, 1976, Queensland Government

Too late to save the unfortunate Noosa estuary, the state government in 1976 issued the report of its Coastal Management Investigation, a document of some four volumes with maps and diagrams. Three of the volumes dealt with details such as wetlands and aquatic vegetation, engineering investigation of canals, and land use investigation.

Commissioned by the Coordinator General's Department in 1974, it required the consulting engineers and surveyors, Messrs Gutteridge, Haskins and Davey, to report on their findings on land use and coastal management in the area all the way from the border of Queensland and N.S.W. to the northern boundary of the Noosa Shire.

The report urges most careful planning of the whole Noosa area, since '. . . it is an outstandingly important resource for its scenic, scientific, educational and recreational value'. It strongly recommends that before any decisions are made on land use that may cause irreversible changes — such as canal estates — further detailed environmental impact studies should be carried out. It states:

The Noosa River and lakes are areas of critical concern to the region. The system is unique for its extent, and for the combination of fresh water and saline conditions of life. Large-scale land development around Lake Weyba, the upper river and the north shore, could have significant effects on the complex tidal, freshwater and saline conditions throughout.

The report says that the yield of fish, crabs and prawns *must* fall away with draining of mangrove and salt marsh areas, as has already happened along the southern bank from the mouth to Tewantin. All seagrass beds, essential as fish-breeding grounds, should be protected from herbicides, pesticides and other pollutants.

"

This was exactly the point made by Dr Arthur Harrold in a submission to the Noosa–Tewantin Chamber of Commerce as early as 1971, when he stated: 'Steps should be taken to secure some of the nursery areas as marine habitat reserves.' A detailed report set out how the development of a canal estate in the mouth would deprive the river system of a significant proportion of mangrove nutrient areas and seagrass beds, already depleted by draining and filling of the south bank.

The C.M.I. Report covered a great deal more than just the Noosa estuary. The area under investigation contained 248 kilometres of open ocean beach, of which forty-two kilometres is in the Noosa Shire. Half of this lies north of the Noosa River. The area south of the river, stretching beyond the shire boundary to Coolum, is the part most heavily used.

Throughout the area covered by the report, long-term erosion from south-easterly drift is predicted to remove up to *one hundred metres of foredune in the next fifty years.* Building on the dunes is now prohibited, and a five-chain erosion prevention area is kept in front of new developments.

The Beach Protection Authority has wide powers in this area, but is reluctant to use them directly. Instead, it encourages local authorities to control such man-made erosion as that from foot traffic and beach buggies, and discharge of storm-water drains.

The Authority deplores the building of rock walls, pointing out that they only protect property and do nothing to restore beaches, while sand-pumping can have harmful effects elsewhere.

The C.M.I. Report recommends that all land adjacent to beaches should be held for public purposes, recreation and conservation, and that all unalienated Crown land should be kept for this purpose, while local authorities should try to acquire coastal land which is freehold. Finally, 'erosion control reserves' should be established along all endangered beaches.

The report includes a detailed assessment of Noosa Heads beach and estuary, after the construction of the Hay's Island development. These were among its findings (author's italics):

The Noosa River–Hay's Island development was constructed by raising the levels of sandbanks on the south side of the Noosa Inlet. When constructed, considerable scour was noticed on the ebb tide in the channel between Noosa Camping Reserve and the development area, together with a deposit of sand on the outer area of the main river channel.

An inspection in May 1974 revealed that the southern canal had now been cut off, with a temporary causeway adjacent to Weyba Creek. Rubble groynes had been erected at the camping reserve and the scour appeared to have been arrested. There are difficulties in estimating current and sediment movement in a tidal canal system.

The estuary appears to be highly sensitive, and it should be studied in detail before further canal estates are approved.

Velocity measurements up to 1.3 metres per second were recorded on an ebb tide in the present main (i.e., southern) channel, compared with 0.6 metres per second in the present subsidiary channel of the Noosa River. The southern channel has developed markedly since the construction of the estate . . . .

Placing gravel or coarse rock on key heads or other beaches subject to wave attack creates an unattractive and almost unusable beach . . . , but it is effective and less expensive than sand nourishment, *if sand cannot simply be obtained by dredging adjacent areas . . . .*

The discharge of treated effluent (sewage) into the canal system direct, may also influence quality of water with regard to nutrients and possible bacterial count . . . .

Dredging in tidal waters can have repercussions: *High turbidity*: Fine silt settles in other areas, kills bottom-living fauna and flora, inhibits plant growth. *Dredging*, though a cheap and attractive source to the developer, causes a disturbed bottom to the stream which may take a long time to regenerate, so that the area is robbed of its living biota . . . . *Removal of sand* from tidal waters represents a nett loss of sand from the estuary-beach system. Where the mouth of the stream is on or close to an ocean beach, this may affect coastal erosion.

Investigation over a long period is essential before leases for such work are granted . . . . Permission should only be given after evaluation of the effect on environment, marine life, etc; it must not affect adversely neighbouring land by changes in the flooding pattern. In sum, canal construction leads to increase in tidal flow, increased velocities, causes bank and bed scour, and could have a major effect on the ecosystems of the area.

Referring to storm surges, the report states that the greatest cyclone influence is in February and March (late summer), generally moving from north-east to south-west. When a tropical cyclone or an intense tropical storm crosses or closely approaches the shore-line, the resultant rise in mean water level is beyond that expected from astronomical tides alone. The rise is caused partly by wind-induced stress piling up a body of water, and partly by lower atmospheric pressure. The rise can be sudden and abrupt.

A cyclone with average winds of only 30 knots and maximum winds of 44 knots could cause tidal surges between three and five metres. All estuaries are subject to storm surges, and the combination of a surge of three metres with a king tide, and floodwaters coming down the river, could be disastrous.

A basic feature of planning coastal zones is that land adjoining all shores should be held in public ownership. The width of esplanades should be at least one hundred metres, and preferably four hundred metres. (The 'esplanade' at Noosa Heads, above the rock wall, was about four metres wide.)

The report urges most careful planning for the Noosa River system because of these factors:

1. The substantially natural conditions of the river and lakes above Tewantin, of Lake Weyba and of the north shore, and the significant changes in the environment that would be caused by large-scale land development.
2. The complex tidal, freshwater and saline conditions in the various lakes and the significant effects that development would have on these conditions.
3. The high productivity of the lakes and rivers in fish, crabs and prawns, which are dependent on the substantial areas of mangroves, salt marsh and seagrass beds, as well as on the bar zone and shallow tidal flats of the estuary.
4. The varied and distinctive vegetation types represented in the adjacent land, and their associated fauna.
5. The value of the river and lakes for most forms of boating recreation.

The report adds that any proposals likely to detract from the scenic, scientific, educational, aesthetic and recreational values of these waters should not be countenanced. Further detailed investigations should be carried out on the Lake Weyba catchment with regard to any proposals for canal development below Tewantin. Canal estates above Tewantin are not recommended at all.

Lake Weyba, being very shallow, is sensitive to pollution. As a fish habitat reserve it is valuable for commercial and recreational fishing, boating, family picnics, and for its beauty. The proposed new coastal highway should therefore not pass to the east of the lake (between it and the coast) where it could seriously affect the recommended national park.

Also, 'any further canal estates should not proceed until more detailed investigations of water flow and siltation, and any effects on the river bar, have been carried out.'

Subsequently the council gave provisional approval for a new canal estate, 'Noosa Waters', to local land developer T. M. Burke Pty Ltd. Their plans were for a multi-million dollar development in the paperbark swamps behind Noosaville. It was stated in the submission to the council that there was no fauna in the area which could be affected.

This was rather too reminiscent of the Cambridge Credit claim that Hay's Island was 'an ecological desert'.

In spite of the urgent warnings of the Coastal Management Investigation Report, the Noosa Shire Council granted approval for subdivision of the canal estate, providing all government departments were satisfied with the proposal.

The Noosa Parks Association is opposed to any further canal developments in the Noosa River and, with the help of the Noosa Midge Control Organization, it decided to contest the issue. The latter believed that the canal banks would provide extended and ideal breeding grounds for Noosa's prime pest, the sandfly.

A Noosa River Protection Committee was formed to organize public protest. The first meeting was held in Tewantin in November 1977.

The meeting was a lively one attended by some two hundred ratepayers, who seemed unanimous in their opposition both to big business in the shape of the developer, and to the council's decision. Three councillors were present.

The deputy chairman, Mr Hassett, contended that council had no option, but was backed into a corner over the matter of giving approval. If they refused the application, he said, the development company could take them to court to show reason for the refusal; and until the impact studies were completed they had no real reasons to give.

Under the Canals Act of 1974, the state government will not carry out such studies until after the local authority has given approval for subdivision of the canal estate. Yet, once this has been given, council has no further right to veto. A further confusion of the issue was that Councillor Peter Sharpe was also a member of T. M. Burke Pty Ltd.

A motion was then put to the meeting by Sir Thomas Hiley, and carried, that the relevant minister be asked to change the Canals Act so as to make it possible for councils to give limited or qualified approval only, final approval being withheld until independent assessments of possible consequences of the canal estate were carried out.

The Noosa Sound development was mentioned frequently by fishermen, among others, who declared that the tidal flow in the top lakes — even in little Lake Como — had changed since the building of the canals, and that they were now finding it hard to make a living. The feeling against Noosa Sound seemed to have hardened into a hostility against all canal developments in Noosa.

A resolution was then passed and a petition, in much the same terms, was circulated for signing. The petition stated:

1. The Noosa River system has deteriorated significantly following the construction of the canal estate 'Noosa Sound'.
2. It can be assumed that the construction of 'Noosa Waters' would cause further deterioration by:
   (a) increasing the likelihood and severity of flooding in the Noosa River, particularly in the Noosaville reaches,
   (b) reducing water quality in the river with consequent deterioration of fisheries,
   (c) aggravating the existing biting-midge problem.
3. Local authority services are already extremely costly to rate-payers and inadequate and could not be extended to another 1000 allotments without additional heavy borrowing.

Your Petitioners, therefore, Humbly Pray that the Parliament of Queensland will:

Refuse to permit the construction of the Noosa Waters development unless:

1. All the recommendations of the Coastal Management Investigation with regard to the Noosa River system are fully implemented;
2. it can be conclusively shown that all the objections outlined in section (2) above will not occur;
3. it can be conclusively shown that there will be no additional burden on ratepayers.

After this no more was heard of the 'Noosa Waters' scheme for some years until, at the end of 1981, new plans were released which prompted the question, 'When is a canal not a canal?' Answer: When it is a lake.

The huge development, covering 150 hectares and providing 1000 new home-sites, now provides for a long, narrow waterway in the shape of a squashed doughnut, with housing both inside and outside the circle. It is claimed that 'unlike canals, the lake will not open up new areas for nature to adjust to, such as sandbanks and channels. . . . The continuous ribbon of water will be connected to the river by a controlled system of inlet and outlet pipes. The waterways . . . will support fish life as well as being a recreational facility. . . . Although the lakes will be replenished regularly they will not be tidal and will not provide a habitat for biting midges.' Or so the developers claim.

So far the new scheme has not been given final approval.

It has been pointed out by the Noosa River Protection Committee that pollution from weedicides, pesticides, fertilizers, garden clippings, and run-off of tar, rubber, petrol and oil from roads, and oil and petrol from motor-boat spillage, will still find its way into the river.

The C.M.I. Report states that, while the construction of each canal estate may cause relatively little change in a river system, the total effect of several such developments may be cumulatively damaging. This is especially so in the Noosa River, where 'the tidal patterns and seasonal variations in freshwater flow appear to be complex, and to have changed significantly in recent years, possibly because of erosion of the beach, or because of the rock wall, or the construction of a canal estate near the river mouth'.

According to the C.M.I. Report, about 60 per cent of the natural vegetation on the mainland south of Tewantin has been destroyed or modified by human activity already. So 'we must preserve adequate samples of natural habitat, as fauna distribution is intimately tied up with vegetation'.

Almost the only remaining areas, whose natural cover is of sufficient extent to support native animals, are in the Noosa–Cooloola district. Therefore, the report recommends the extension of the Cooloola National Park southwards to link up with available areas on the north shore of the Noosa River, incorporating almost the whole of the eastern shores of Lakes Cootharaba and Cooroibah into an enlarged park.

One whole section of the findings of the coastal management investigators (Mr Peter Waterman and Ms Stella Sabljak) dealing with 'case studies of recreational land use and availability of public space at Noosa and Caloundra' seems to have been suppressed; or, as an official of the Coordinator General's Department put it, 'We didn't act upon it.' It was not issued along with the other findings to the Steering Committee appointed by the councils involved.

I have it on good authority that the copies circulated to the various heads of departments in the Lands Administration were all called in and destroyed. One, however, remained in the Coordinator General's library, and I have obtained a copy of this. Even before the findings were issued, according to one of the investigators, 'heavy pressure' was put upon them not to offend big business in the form of land developers, especially of canal estates. The pressure came from the top downwards — from the Treasurer's Department of the Queensland government.

Yet all the Treasury stood to gain from the agreement over the Noosa Sound development was a rental of $800 a year (Development Lease No. 8, under the agreement with the lessee). In addition, all dredged material used for reclamation was subject to a royalty of 7½ cents per cubic yard; and 10 per cent of the gross sum for which each block was sold went to the Crown Lands Department, in return for a Deed of Grant for that block.

At least 8 per cent of the development was to be surrendered to the Crown for park and recreation purposes, 18 per cent of the water frontage had to be kept for public access, and beaches 'must be established on the park frontages to the satisfaction of the Minister for Lands, the Department of Harbours and Marine, and the Noosa Shire Council'.

A bond for $100 000 had to be lodged with the minister as surety for the completion of the work: and another bond of $70 000 with the Noosa Shire Council against the cost of a bridge to Munna Point, half to be borne by council.

Summing up, the investigators found that:

1. Areas allocated for public open space are inadequate and poorly located to meet the needs of either permanent residents or tourists. An allocation of only five per cent of a subdivision to public open space is a gross underestimate of the need. The practice of local government authorities accepting money (in Noosa Shire, $20.00 per allotment) as an alternative to public space, must be stopped.
2. Townscapes exhibit clearly the historical pattern of subdivision without regard to the topography of the area . . . . This condition was examined in detail at Sunshine Beach, Noosa Heads, and it was concluded that *the poor physical planning could only be rectified by resumption and re-allocation.*
3. National park areas near Noosa should be enlarged, consolidated and increased . . . .
   Provision should be made for the incorporation of naturally vegetated areas into townscapes as alternatives to the cleared and sterile picnic/playgrounds . . . .
   Development of housing and roads has taken place too close to high-water mark, so the land with the greatest recreational potential is denied to the public.
4. **Wildflower Reserves:** Relatively small areas of land (20 perches) can suffice to provide protection for a species, so the opportunities for creating small reserves are many . . . .
   Small wildflower reserves scattered through the area would also provide much needed open space.
5. The Noosa Parks Association is seeking national park status for the timber reserve (T.R.977) between Lake Weyba and the Noosa Coastal Highway. This area contains a profusion of wildflowers: 175 species were recorded in 1969. Grey kangaroos, swamp wallabies and echidnas are reasonably common; swamp pheasants and other birds are numerous. This would make an assured recreational outlet in the vicinity of the rapidly growing townships of Coolum and Peregian. It could help to eliminate some of the ribbon development of the coastal strip from Noosa to Caloundra.
6. There is need for increased open space to serve the requirements of Noosa Shire alone. The need is greater still if the full requirements of the expansion of population on the Sunshine Coast are taken into consideration.

The investigators recommended other additions to the national park:

(a) The two hectare quarry reserve (R.874) which has never been worked, next to the existing (abandoned) quarry.
(b) Portion 29, an area of eight hectares immediately to the north of R.874. [Became Special Lease 2945 in 1964, owned by chairman of Noosa Shire Council, Ian Macdonald.] The southern portion is rocky and precipitous and unsuitable for subdivision.
(c) Portion 1653, south of the Noosa National Park, and west of Noosa Coastal Highway. [See parish map 1931. This shows the 382 hectares as belonging to C. S. Miles, with six hectares at the top corner as a 'proposed recreation reserve'. This is where the Noosa Bowling Club and tennis courts are now. The only high land at the north end has been subdivided by council, which now holds the freehold for the whole Portion, and sold as Cooloola Estate. The rest is low-lying peat swamp surrounded by paperbarks.]

It is strongly recommended that this portion be added to the national park and be used as a flood plain for Sunrise Estates. The present drainage scheme for this estate, across the dunes to the beach, has proved disastrous. Scouring caused by the drainage outlets is well-documented.

In the case study for Sunshine Beach, the findings were:

1. **Sunshine Beach - Sunrise Estates.** Sunshine Beach . . . has no parkland large enough to be used as a playing field, a community centre or for a surf lifesaving club. Subdivisions have been progressively effected by T. M. Burke Pty Ltd without any provision for public utilities or space to provide some visual relief on the landscape.
   . . . . *All* the 'parks' in the Sunshine Beach area are in some way connected to drainage basins and none of them can be used as parks for recreational purposes.
   At the township itself, a park of approximately one hectare exists near the northern extremity of the township. The park is a gully, unrecognizable as a park, and unusable for any purpose. The only flat portion houses a row of toilets and a car park.

   The one hectare area of land designated as park at Belmore Terrace is another public park heavily scoured by storm water so that its area has been reduced significantly. Parking facilities are inadequate on weekends and cars park along the esplanade.

2. **The Parade, Noosa Heads.** The strip of beachfront running parallel to Hastings Street . . . is, in effect, a private beach for the motels and holiday units bordering it. There are only two access points, one at either end of Hastings Street.

There are only three local government caravan and camping reserves within the Noosa area. [Now only two, as the Tewantin river-front park has been coopted for the new shire offices.]

3. **Little Cove Road.** Little Cove Road is used as a parking lot by visitors to the beach downhill from Laguna Hill. As the road is very narrow, the parked cars create a considerable traffic hazard and a nuisance to local residents; again illustrating that inadequate land has been provided for public use.

'As it was, shall be . . . .' The sandspit protecting the lagoon behind Noosa Woods in the 1950s. This was a great place for whiting up to ten years ago. (By courtesy Kevin Freeman.)

Work in progress on the new river mouth at Noosa, which now cuts off the last bend and issues hundreds of metres north of the former entrance. (By courtesy the Noosa News.)

11 What's done cannot be undone

The aesthetic qualities of the Noosa River are outstanding. No visitor to the river can help but be struck by its beauty. It could easily be argued that the whole system should be protected for this reason alone . . . . Because of their uniqueness, the features of the Noosa River should remain public and not be exploited by a privileged few.

S. Sabljak and P. Waterman, *Coastal Management Investigation Report*, August 1974, Queensland Government

After a look at the history of the Noosa estuary and its beach in the last hundred years, what seems to be established is that it has always been vulnerable. The beach has washed away many times, exposing the underlying rock, and come back unaided. The bar and entrance have moved up and down, cutting away first the south-east point, then the north-west. (There is really only one 'Head' at Noosa; the far side of the entrance is no more than a low, shifting sandspit.) More recently, in the year 1977–78, the swing was to the north again.

Taking advantage of this movement, and two summers of calm weather, hydraulic engineers firmly pushed the mouth of the river even further towards the north. They extended the south-east spit, beyond Noosa Woods, as far again as the whole length of the old Noosa Heads beach (restoring it to where it was up to ten years ago). Through a massive sand-pumping operation, this spit was raised to a four-metre sandhill, which was also extended the length of the beach, completely covering the controversial rock wall.

At the enormous cost of $1 400 000, the unstable river mouth has been fixed with a rock revetment and groyne, the river has been induced once more to flow through its winding northern channel, and the Noosa Sound canal estate has been protected from the open sea.

The immediate benefit was a wide, new beach of rather dirty, greyish sand at Noosa Heads, extending twice the length of the original beach.

Advocates of the scheme had scarcely finished congratulating themselves on the 'favourable outcome' before the main beach started to disappear once more. After some heavy but not cyclonic seas, the beach developed a three-metre cliff of sand, dangerous to pedestrians and to swimmers at high tide. By 1980 there was only the narrow esplanade of sand at the top of a bare rock wall. Some high tides with north-easterly winds soon removed the last vestige of the artificially pumped sand; it was back to square one.

Some large new developments next to Los Nidos, where there was less erosion because the natural curve of the beach is retained there, pushed some new sand above the rocks from their building excavations. Otherwise the main beach below the expensive motels is worse than before. Surfboard riders were injured against the rocks during a recent high-tide rescue, and a dangerous rip has developed.

During this steady deterioration, the Beach Protection Authority, which had approved the restoration scheme, kept issuing bulletins to the effect that the beach would restore itself 'when conditions were right'. As conditions were better than at any other time in the past five years, with a huge build-up of sand along Sunshine and Peregian Beaches, this could only be described as whistling in the dark.

However, the sandy spit inside the rock groyne made a new beach a kilometre further north, from which catamarans with their rainbow sails could be launched and where swimmers could sunbathe, while wind-surfers plied on the newly sheltered inlet. The new dune was stabilized with a thick growth of casuarina and coastal myall; but it must be remembered that the original spit washed away in the 1967–68 floods was covered with old casuarinas of large growth, which did not save it.

The shire is now in debt to the tune of $400 000, and will be responsible for the future upkeep of the revetment, groyne and sand-dune. No one knows whether the groyne will stand up to floods and cyclones, or how a straighter and deeper outlet will affect the river in the long term. It was reported that fishing boats were leaving the river, as the bar could not be negotiated at most times; sailing boats and outboards were going aground where there used to be a clear channel; and there were no fish or prawns to be found anywhere.

All this may stabilize in time; or a succession of cyclones could sweep the rock groyne away and leave the river free to slice through the expensive artificial dune. Training walls have worked well at Mooloolaba, further down the coast, because the

October 1978. The rock groyne in place. (By courtesy the Noosa News.*)*

October 1981: the re-sited river mouth from the air. The entrance has moved further north, and has a shallow, silted channel, while the north head has lost its protective sandspit and many trees to the encroaching sea.

Compare with the vertical aerial photograph in 1978 (colour plate), when the deep channel was against the rock training wall. (Photo: Bill Griffiths.)

Mooloolah river mouth is directly sheltered by Point Cartwright. At Noosa Heads the mouth is nearly two kilometres from the headland.

The engineers who designed the boulder wall at Noosa Heads say that this is all that saved Hastings Street in the last bad cyclone year of 1976. However, the first alarm about 'losing the beach' occurred in 1928, and most of Noosa Heads is still there after fifty more years of tropical storms.

In the last ten years, however, things have begun to change rapidly and alarmingly. The fragile sandspit, denuded of most of its rainforest and beach flora protection, its fringing banksias and casuarinas, has become a piece of fashionable resort land jammed with millions of dollars' worth of investment. An artificial island estate has been raised in the river inlet, protected with rocks round its perimeter. The rock wall built in 1968–69 was extended right round the end of Noosa Woods.

Experts, who were not consulted when the wall was built, pointed out that the slowly subsiding wall could fail in a bad cyclone with tidal surge; and the canal estate was at risk.

As long ago as 1970, Professor R. Langford-Smith of the University of Sydney wrote:

The ocean front properties should never have been allowed to encroach on the foredune, let alone the berm [the level area below] itself. . . . Given time, however, I have no doubt that Noosa beach would have recovered fully if left alone. . . . Our experience in N.S.W. is that a sea wall built in this type of environment . . . defeats its own purpose, and I can see no hope of Noosa ever recovering to its former state . . . .

After 1976, everyone agreed that 'something' had to be done about the deterioration of the beach and the threat to Noosa Sound; the million dollar question was — what?

The latest study of the Noosa River and its problems was made by Professor G. McKay, of the Department of Civil Engineering at Queensland University. He set up a hydrographic model of the whole river system; but he was baffled by the complexity of these waterways with their slow welling of rainfall seeping through giant sandhills, and their chain of tidal lakes. Because of the low rate of fall from the highest lake to sea level, he foresaw a danger in any scheme to alter and deepen the mouth of the river, which might cause the smaller lakes to drain away.

He put forward three alternative suggestions to the state government, the cheapest to cost $800 000, the dearest $1 200 000. This last plan called for the stockpiling of masses of sand, to be pumped from the northern spit and inside the estuary sand-islands, using very little rock. State cabinet and the Noosa Shire Council agreed to go ahead with this scheme.

Professor McKay stressed to me that it was necessary to proceed slowly with each stage. It would be a matter of resiting sand already there. Beach sand, he said, should *never* be removed for fill on building sites or other purposes, as it becomes lost forever to the beach system.

Dredging in the river mouth and at Munna Point had deepened the channels in the Noosa and altered the speed of flow. There was now a conflict of interests between fishermen, tourists, and conservationists, all of whom want to see the natural river mouth preserved, and property owners at the Sound who want their property protected at all costs.

He added that another hazard for the Noosa estuary was the effect of more and more subdivisions for housing in the catchment area. Engineers were only just beginning to realize the effect of roads and buildings on the annual flow of a river such as the Noosa, where the whole catchment area was sand that absorbed heavy rainfall like a giant sponge, and released it slowly and uniformly to the river:

But when you get clearing, and still more, building with non-absorbent roofs and driveways, roads and parking areas, the speed of run-off in heavy rain increases dramatically. This first became noticeable in the floods of early 1968, when the Point at the Woods began to erode, and the sheltered lagoons were gradually lost.

The rescue plan for Noosa Heads was drawn up by the Beach Protection Authority, based on Professor McKay's recommendations. However, the professor feels that a sandbank alone should have been placed experimentally at the south-east of the river mouth, rather than a rock groyne. Then, if it didn't work, it could be pulled out easily (since no one knows what effect a wall might have on the upper reaches of the river). Even if a bank of sand should get washed away, it would not cause any harm to the beach; whereas a rock wall that failed, or was found to have a detrimental effect on the river, could be disastrous.

The scheme means that residents on Noosa Sound will be subjected to a 'benefitted persons' tax', or extra rates, to pay for upkeep of the new works, which will be a council responsibility in the future. Residents on the seafront at Hastings Street, who already pay higher rates, now have only a rock wall between them and the sea. As one ratepayer wrote angrily to the *Noosa News*:

Who would in his wildest fancies ever have imagined that we would have a council that would destroy Noosa beach, our prime holiday asset, for the benefit of twenty-six owners who choose to build on the seaward side of a sandspit facing the Pacific Ocean; ensuring moreover that not one of them would lose a square foot of ground? . . . The beach must be restored, and no doubt we must pay for it . . . .

He blamed the rock wall for the beach's loss; and it was approved by the same Beach Protection Authority. One trouble with the B.P.A. is that it has no money to persuade, nor power to coerce, local councils into following its recommendations. It is something of a toothless tiger. As for the Department of Harbours and Marine, it has acted like Pontius Pilate over the Noosa Sound and Noosa Waters canal estates, washing its hands of responsibility. The council itself seems to be floundering in its decisions, through the combined handicaps of lack of specialized knowledge, not having its own resident engineer, and through pressures put upon it, however subtly, by large vested interests.

An overall coastal authority for each state has been suggested, both by the Coastal Management Investigation Report and the Australian Conservation Foundation Symposium on The Australian Coast (1969): 'The creation of powerful planning authorities at state level to coordinate planning for the whole coast zone and its resources for the long-term benefit of all Australians, rather than for the benefit chiefly of local and other vested interests.'* Special Commonwealth funds should be allocated to help state planning authorities in protection work for the coast, 'which cannot be carried out on an *ad hoc* basis by local government with very restricted financial resources'.†

Noosa Heads and the Noosa estuary provide an awful example of what can happen to a lovely coastal resort which has been opened up without any regard for master-planning of the whole area. Noosa is a particularly fragile environment because so much of it is 'a house builded on sand': 'And the rains came, and beat upon that house, and it fell; and great was the fall thereof.'

Clear-felling of wallum scrub, rainforest,

---

*Special Publication No. 7, Australian Conservation Foundation, Melbourne, 1972.

†Ibid.

cypress pine and paperbark stands has gone on all over the Noosa area, except for the North Shore and the forestry reserves. Many of the latter are now being cleared for slash pine plantings. Clearing is dangerous on steep sandhills, such as those on which much of Noosa Heads and Sunshine Beach is built. The council's workmen themselves bulldozed nearly every tree on Cooloola Hill when the first section was being developed for sale, with bitumen roads and kerbing. Such is the rainfall and the regenerative power of the local bush, that the hill managed to cover itself with regrowth before building began.

Much clearing has been done by individuals. New arrivals who come from one of the big cities in the south, attracted by the greenery of Noosa, proceed to turn their twenty perches into a neat copy of the cement-drive-and-lawn suburban block they have left behind. Councillors say that a by-law preventing an owner from cutting down trees on his block would destroy the rights of the individual. In spite of this, a council by-law already requires him to clear his block of brush if it should be a bushfire hazard, and of noxious weeds.

In about 1972, a large section of sand-dune country behind the national park was clear-felled, leaving an eroding sandy space known as 'the moonscape'. Permission was not given for subdivision till 1977, when regrowth had started to heal this vast scar.

The developer's 'ponding tank' gave way in heavy rains and part of the Sunshine Beach recreation area disappeared down a new gulch, while a bitumen road lost a section to the beach and a fence was left hanging in mid-air.

At Sunrise Estate, elaborate cement pipes for drainage and steps for access to the beach both washed away, leaving a jumble of smashed concrete and a deep sandy gully. Some run-off from new estates has been channelled to a natural freshwater lagoon behind the frontal dune, and this has sometimes broken through in flood time. The pure, sandy soils have little resistance once the narrow layer of humus is removed. The Beach Protection Authority is trying to direct foot-

Sheltered lagoon behind Noosa Woods, re-created by pumping the new beach. The yacht Laivinya, *in foreground, is from New Zealand. Behind is the new housing estate and rock perimeter of Noosa Sound. (Photo: Mike Noone.)*

traffic, by means of paths and brush fences (even barbed wire, which becomes dangerous when half buried in sand) to use regular routes and not tread down the natural grasses and wildflowers.

It is expected that even the North Shore will be urbanized in thirty years' time; and 3000 people will live at Sunshine Beach, which is still a village.

None of the newcomers will be able to buy seafront land; the days of such development are almost over. However, campers cut down trees for firewood, scatter their indestructible plastic rubbish, run over the eugarie beds and break down the dunes on the remote North Shore today. The beach has become as busy as a highway; even traffic accidents occur there. The ghastly pattern is being repeated further and further afield. The rubbish and the human excrement are spreading right up to the top of Fraser Island.

Apart from development, weekend tourism is exerting a growing pressure on the Sunshine Coast. The increase in incomes and in private car and boat ownership, more leisure time, more caravans, the growing popularity of camping and fishing all mean that hundreds of thousands of city people set off for the coast by road each year. Picnickers unthinkingly trample down the sea-verge wildflowers and break the crests of dunes, scattering their glass and plastic rubbish, their milk cartons and newspapers.

This once remote and beautiful area, populated by a few fishermen and farmers, has deteriorated visually and aesthetically. However it is not only tourism that is to blame. It is our own apathy, our selfishness and greed, our indifference to what is happening to our environment, that have led to the rape of Noosa.

The rainfall, the incidence of cyclones, have not changed with the years. We have changed the conditions simply by being there.

The beach has been changed beyond recognition. At one end it looks like the Sahara, at the other it has eroded back to bare rock once more.

It is too late now to undo what has been done. Only by tipping more money into what one councillor has described as 'a bottomless pit' can we hope to save the expensive real estate and perhaps recover some semblance of natural beach.

The council's most recent plan, for which tenders are being called, is for yet more rocks — $400 000 worth — to be placed in a groyne at a right angle from Noosa Woods. 'Experts' are predicting that this will create sand build-up and will restore the beach — with the help of pumped sand, which will cost an estimated $250 000 extra.

The latest threat to the battered Noosa estuary (apart from the continuing threat of flood and cyclone each wet season), and the latest development scheme to raise its ugly head, is the proposal for a deep-water marina. The plan was suggested by the Lands Commission, which has asked for residents' opinions. The marina would be built in the river waters sheltered by the new sandspit, the fragile barrier between the river and the sea created in 1979. For once, the council is not involved, having asked for the reclaimed area to be zoned 'open space and recreational', and transferred to the council.

The proponents of the marina have actually suggested digging a deep channel through the beach from the open sea to the sheltered water, thus, of course, destroying part of the man-made barrier erected at such cost. The channel, protected by rock walls, would cut right across the new section of beach, and prevent people walking along it to the mouth. The popular swimming, fishing and wind-surfing activities in the sheltered waters would be crowded out by boat anchorages and jetties. The arguments for and against have been bitter.

In 1982, with cyclones hovering off the coast and sending in some heavy south-east swells (though so far away), the sea appears to be deciding the outcome of what the local press has called 'The Great Debate'. The sand at the far end of the rock wall has been cut away to form a dangerous four-metre cliff, and the first line of replanted casuarinas has fallen in. Several posts of the bordering fence fell over the cliff in one night. Without any flooding in the river, without any gale-force winds or tidal surge, the far strip of beach seems to be following the first. Once gone, it could leave the rock groyne isolated, a dangerous island that would soon disintegrate: the disaster warned of by Professor McKay, 'a rock wall that failed'.

Fifty years ago it was possible to remove the only endangered property at Noosa Heads with sixty men and a draught horse. If the problem is vastly greater now, it is only because, through lack of planning and foresight, we have allowed building where it should not have been allowed.

Today we can only look back and mourn the passing of that 'picturesque beach' which Andrew Petrie and his companions saw in all its pristine beauty only a hundred and forty years ago.

12 Reaping the whirlwind

The 'multimillion dollar investment' is no longer a rarity now that big business has discovered Noosa Heads and its 'new', remarkable, disappearing beach. In one fortnight in March 1979 two new plans were announced as being on the drawing board, and being considered by council:

A $20 million 'holiday and sports complex', with hotel and convention centre, each of 16 storeys, complete with marina, golf courses and bowling greens — to cover 36 hectares of undulating banksia scrub overlooking the river and Noosa Sound. This area has been 'selectively cleared' but the project has not been proceeded with.

A $100 million design for changing Hastings Street into a fashionable, expensive resort like the suburb of Waikiki in Honolulu. A Hastings Street motel proprietor (who just happened to be a shire councillor) had been quietly organizing neighbouring property owners into a group, and brought in architects, consultants and 'experts' from Hawaii. The plan included a $12 million multistorey building.

As anyone knows who has seen the overcrowded foreshore of the overrated Waikiki Beach, with the Waikiki Hilton sticking out into the sea at high tide and shadowing the beach with its 30 storeys, this is an unplanned and undisciplined disaster area. (Compare it with, say, the tasteful French colonial style of Papeete in Tahiti.)

American planners and designers from Hawaii are the last people we should consult: residents there are desperately trying to stop the cancer of high-rise buildings from spreading to the windward side of Oahu and the relatively unspoilt outer islands. Yet Maui already has its million-dollar 'sports complexes' and high-rise hotels on the beachfront.

When Christmas 1980 and the traditional 'full to the brim' period of August/September showed that many units and motels had vacancy signs up, local businessmen began to be worried. Small businesses were suffering too from the lack of tourists — hire-boat operators, take-away food shops, and smaller supermarkets now feeling the competition of the giant Noosa Fair. A big promotion programme to attract tourists, and to assure them that Noosa was not only for the rich, was begun in the south and even extended overseas to Singapore and Japan.

By Christmas 1981 things were nearly back to normal, but the fact remained that the family groups who used to camp or stay in caravans, or rent accommodation for several weeks, were no longer coming. At least four caravan parks had closed, to be covered with more lucrative 'unit development', while the popular shire caravan park had been ousted by the new council chambers. There was an oversupply of 'luxury' accommodation.

There was a mad rush to build this sort of accommodation and cash in on the high rents. Now, from Peregian north to Noosa Sound, there are so many that they cannot all be filled, and applications for special building permits are falling off after reaching astronomical numbers in 1979–80. Yet units are still being sold 'off the plan', before they are even built, to eager speculators looking for a quick capital gain. When finished and furnished they can be sold after twelve months for a tax-free profit of around 25 per cent. They are not being built to live in, or even to rent, but to make money in the inflationary spiral. According to a report from the National Tourist Industry Data Centre, the demand is still outstripping the supply, and unit turnover, including new sales and resales, had reached $48 million for the year to January 1981.

Noosa, meaning in effect Noosa Heads, had been getting some bad publicity in the southern press from disillusioned visitors who did not like what was happening there, who reported that the 'beach' was lined with rocks and the prices in Hastings Street were unreasonably high. One writer referred to it as 'Not St Tropez — more like Gosford, with pretensions.'

This of course was unfair, as nothing so far has altered Noosa's beautiful climate in winter, said to be among the best in the world. It brings settlers from as far afield as South Africa and New Guinea, England and Canada, as well as refugees from the cold of Canberra and Melbourne who are flocking to the north to live.

There were some fears that Noosa Heads' unique micro-climate might have been affected by the wholesale clearing of trees. There were two very dry and dusty summers, with yearly rainfall averages down to less than 1000 mm, but the summer of 1981–82 has brought the normal heavy monsoon rains, with severe thunderstorms and even hail.

Some journalists, reacting to the over-promotion of the Noosa image, came to see for themselves and returned to criticize. The shire council actually proposed sueing Graham Williams and the *Sydney Morning Herald* for saying in the course of two articles that Noosa 'had no beach'. Under the heading 'Lovely Noosa: Is the Final Chapter Developing?' he described the waves pounding on the rock wall below the expensive units (as they do at high tide). He then detailed the $100 million project of the Noosa Resort Corporation for the 'planned development' of Hastings Street. This group already owned half the valuable real estate there: Thatcher's caravan park, the Pine Trees chain of motels and restaurants, and a large parcel of vacant land between Hastings Street and Noosa Sound, besides part of Motel Hill and of Stage Three of the Noosa Sound project.

They talked a great deal of a master plan for the area, including a $12 million, six-storey development, but their ideas were described by at least one shire councillor as 'grandiose and vague'. It was rumoured that they were having some 'liquidity problems'.

Then the Local Government Minister, Mr Russ Hinze, came to Noosa Heads for a meeting with the Noosa Shire Council, ostensibly under one of his other hats as Main Roads Minister. He proceeded to let the cat out of the bag: Noosa Resort had asked for backing of $7 million from the Queensland government; this was being considered, but first he would like to sound out the opinion of council on the granting of a franchise to the developers. This would bypass council controls and by-laws, but the developers would have to provide their own infrastructure of water, roads, sewerage, etc.

Though asked to keep the proposal 'under wraps', two councillors released the story to the local press, and Noosa ratepayers were immediately up in arms. They didn't want an Iwasaki-type franchise in their town.

Hinze was furious and said the councillors had 'betrayed his confidence'. A public meeting to discuss the franchise proposal was called. It overflowed the hall, with five hundred inside and a hundred or more outside in the corridors. No representative of Noosa Resort appeared. The fact that the Resort Corporation still owed its last year's rates, besides several other debts running into thousands of dollars, was cited. The meeting was overwhelmingly hostile to the proposal, and a Noosa Action Committee was formed to oppose it.

Since then, the Corporation has become quieter and quieter.

First it sold the caravan park for a reputed $4 million, but its mortgagors said it had no right to sell at auction; finally a private sale was negotiated with Kern Bros. Next there were rumours that its joint-venture partner, Adelaide Holdings (in debt to the order of $19 million), was planning to sell the Munna Point land. Next, 'as part of a general programme of consolidation', the Corporation announced it would sell another prime piece of Hastings Street real estate, more than 5000 square metres of land for which it hoped to receive a bid of $6 million. Meanwhile the government announced that, 'because of the lack of corporate structure', the Corporation would not be receiving a financial guarantee.

The Munna Point land (Stage Three of the Noosa Sound development) is now in the hands of receivers in what is almost a replay of the situation in 1974 when Cambridge Credit ran out of funds.

As for the 'master plan' for Hastings Street, Noosa Resort no longer owns or controls most of the land involved. The team of alert new councillors appointed since the March elections ('The Residents' Team') pointed out to the Town Planning and Building Committee that the previous council's approval of the plan (which included the relaxation of several town plan by-laws and requirements for building heights and site coverage) was therefore no longer valid.

However, Noosa Resort Corporation had put in an application for one of its Hastings Street projects, a building advertised as 'six storeys over car-parking' but which, in fact, was nearer to nine storeys with some car-parking area half underground. The committee recommended against approval, as the building would exceed the council's height limit of twenty metres. The refusal was made official by an almost unanimous vote of the full council in May 1982.

The resolution passed was 'That Council resolves to refuse TPC 1530 for Portion 16 Hastings Street, Noosa Heads, on the grounds that the application does not comply with Council's by-laws and policies.'

As Councillor Noel Playford pointed out, 'The worst feature of the whole exercise was that the council was asked to give approval to a master plan for a large part of Hastings Street, a plan that had not been made public. Therefore, the people had no chance to comment on this major development proposal.'

It is only fifteen years since a large piece of absolute beachfront in Hastings Street could be bought for $6000. The present prices seem quite unreal: the owner of one of the controversial First Point luxury units recently refused a million dollars for his apartment.

A few older residents cling grimly to their old homes on the seafront, surrounded by the noise of cranes and pile-drivers, and refuse the most tempting million-dollar offers.

One of the results of the overdevelopment and unrealistic prices in Noosa Heads is that the hinterland has become more popular and hence more valuable. Even in sleepy Tewantin, small building blocks fetch $20 000–$25 000. And the latest thing is the 'country estate', which is subdivided into large blocks, retaining as much as possible of the natural tree cover, and sometimes incorporating a golf course. The Noosa Parklands development has proved the popularity of this uncrowded style of living. There are now Noosa River Heights, Noosa Valley estate and the Noosa Outlook estate, all offering an alternative to the heaps of concrete boxes going up so rapidly at Sunshine Beach and Noosa Sound.

The latest architectural gimmick (after brick arches, which are 'in') is to decorate balconies with artificial Italian marble balustrades (made of concrete)! These are stuck on the front of red-brick suburban villas in an attempt to make them look 'Mediterranean'. Instead, they stick out like a row of ill-fitting plastic teeth. And one of the new 'plazas' actually has plans for an indoor aviary — in an area where the native birds, almost tame in the flowering trees, have always been a natural attraction.

Although, with a self-congratulatory fanfare, the council introduced a tree-preserving by-law in 1979, by which it became an offence to remove a tree with a diameter of fifty centimetres or more at the trunk, this has proved completely useless. Building plans are passed without any inspection of the site for existing trees, and once the developer has his building permit he is free to cut down everything on the block with impunity, and everything within several metres outside it that interferes with his building. Therefore he does so, as it is easier to bulldoze the whole area back to bare sand than try to avoid the trees already there. And a few 'instant palms' will soon add an air of spurious greenery to the front elevation to match the architects' inevitably leafy and embowered drawings.

With the present Works and Services Committee, indeed, there seems to be a positive policy of destruction.

Residents of Peregian Beach were indignant to find that a regenerated sandhill which had formed on the beachfront around two casuarina trees, and covered itself with a natural mat of interwoven beach grasses, convolvulus and pig-face, had been bulldozed out of sight and flattened, 'to prevent erosion'! Visitors began walking over this bare area to the beach, so the Works Committee then erected a large, expensive and ugly barbed-wire enclosure and put up a notice, 'Beach erosion area'. There had been no erosion before this wanton act.

The Parks and Gardens Department, behaving with all the rationality of a hen with its head chopped off, proceeded to plant exotic coconut palms, which they then left to die, as water was rationed. On Noosa Sound they cut down the re-established acacias, native to the area, 'in order to plant better trees'. They then installed a row of tiny fifteen-centimetre seedlings down the median strip and left them to take their chance in scorching salt winds.

In the case of the old fig-tree cut down for Kern Bros' Hastings Street development, the council by-law was useless. Finally, to show the utter hollowness of its attitude on tree-preservation, the council itself voted for the removal of three magnificent scribbly-gums, a hundred and fifty years old, from the roadside of a new development at Tewantin, though thirty-six of the thirty-eight residents in the street wanted them to remain. The developers and the council's own engineers had agreed to a slight deviation in the road to save the trees. 'Expert advice' was called in to say the trees would die anyway — though they would normally live four or five hundred years. The trees came down.

'People are the predators who destroy places,' says 76-year-old Harry Spring, of the original

Noosa Heads main beach, December 1981, low tide.
The 'new beach' of pumped sand has come and gone.
(Photo: Mike Noone.)

136

The huge Moreton Bay fig, overtopping a two-storey building and with a canopy 15 metres in diameter, which was cut down by Kern Bros for their Hastings Street development, 1981. (Photo: Mike Noone.)

'Harry Spring's hut', which is a popular destination for cruises and picnic trips up-river. 'If Noosa had been kept the way it was, today we would be able to put a cover charge on people wanting to see it.' He says that greed and speculation have turned parts of Noosa into 'an Alcatraz', and that a capital-gains tax is the only way to bring back sanity.

'Luxury' is the latest word in the real estate agents' lexicon. A 'luxury unit' used to mean that it was equipped with such things as dishwashers, automatic washing machines and dryers. Now it is gold-plated taps, imported marble floors, velvet-lined walls, and spa baths in every *en suite* bathroom; and outside, of course, palm trees. Noosa is being marketed as a tropical paradise of waving palms, which it never was.

It should be remembered that the Latin word 'luxus' means, besides luxury, excess, extravagance, redundancy. The latest Little Cove development, opposite First Point, is to have individual swimming pools and saunas, imported marble floors, and artificial chlorinated streams. Each unit will cost at least $500 000.

The Noosa Resort Corporation's plan for turning Hastings Street into a fashionable, overcrowded Waikiki, and the other scheme for a $20 million holiday and sports complex, including marina, are still being floated. Already approved is a multi-million dollar tourist resort on the steep hillside above Morwong Drive, with chairlifts to carry tired businessmen up there, and tennis courts suspended in mid-air. Anything is possible — including paving the whole of Laguna Bay foreshore with concrete — if enough money is forthcoming.

But as one long-time resident observed, 'Let them wait for a return to some of the cyclonic weather of years ago. Their eyes will be sticking out like sand-crabs'!'

For the new beach that suddenly attracted all these large investments has yet to be tested in a cyclone or a flood. If the river cuts through to its old channel, does the shire spend another quarter of a million dollars on protecting the new marina as it has already in saving Noosa Sound?

And with all our present water-supply problems — reduced last winter to watering lawns with a bucket — our sewage problems and what the Beach Protection Authority calls a 'lack-of-beach situation', do we really want canal and hotel and marina developments catering for another 3000–5000 people? As it is, we don't have enough water for our gardens. Then we have the prospect of massive quantities of insecticides and garden fertilizer draining into a marine nursery.

Surely the developers with their millions of dollars — some reputed to come from Hong Kong, Singapore and the Middle East — should be told to go elsewhere, and not use their money to destroy such a delicately balanced and fragile ecology as that of the Noosa estuary.

Appendix A

Appendix I to the Coastal Management Study, Vol. 2: Species of birds reported in the investigation area.

List I. Birds recorded from mangroves:

rufous fantail
mangrove warbler
mangrove honeyeater
mangrove kingfisher
marsh crake
lotus bird
white-headed stilt
banded stilt
glossy ibis
white ibis
straw-necked ibis
little egret
white-faced heron
black cormorant
little black cormorant
white-headed shelduck
mangrove bittern
black bittern
grey-tailed tattler
reef heron
red-crowned pigeon
red-backed sea-eagle
white-breasted sea-eagle
silvereye
goshawk
brown honeyeater
yellow-faced honeyeater
white-cheeked honeyeater
grey fantail
brown thornbill

List 2.

A. *Migratory birds found on estuarine sandflats:*

grey-crowned knot
stripe-crowned knot
red-capped dotterel
Mongolian sand-dotterel
grey-tailed tattler
greenshank
common sandpiper
sharp-tailed sandpiper
golden plover
bar-tailed godwit
black-tailed godwit
eastern curlew
whimbrel
little whimbrel
broad-billed sandpiper

B. *Non-migratory birds found on estuarine sandflats:*
Caspian tern
crested tern
little tern
silver gull
pied oystercatcher
pelican
banded stilt
yellow-billed spoonbill
white-faced heron
black cormorant
little pied cormorant

List 3. Migratory birds found on salt marshes:

oriental dotterel
double-banded dotterel
terek sandpiper
greenshank
red-necked stint
curlew sandpiper
grey plover
Japanese snipe

It is to be noted that lists are not exhaustive but show the common species . . . .

(*Author's Note:* Many of the non-migratory birds are still to be found in other parts of the estuary: terns, pelicans, stilts, gulls, herons, sea eagles, ibis, egrets and cormorants. But of the migratory birds in lists 2 and 3, most seem to have disappeared from the Noosa estuary along with the mangrove inhabitants.)

Appendix B

From: Case studies of recreational land use and availability of public space at Noosa and Caloundra: *Coastal Management Investigation Report*, August 1974. Commissioned by the Queensland state government (author's italics).

Noosa Heads area has sufficient vacant, subdivided land (57 per cent of all allotments within the town site) to facilitate a replanning . . . to meet demands of the increased regional population to the year 2010. Replanning must take into account the poorly planned commercial centre in Hastings Street and restricted public access to the beach along the Parade.

Within the study area there is often a visual appearance of adequate bushland and undeveloped space. This . . . however, is a temporary illusion as the existing vegetated stands have been alienated and subdivided. When development of these areas occurs, *over 90 per cent of the vegetation within the townscapes* (Sunshine Beach, Noosa, Noosaville and Tewantin) *will have been removed.* This reduces the available flora habitats and can seriously affect:

    (i)   ground water-table conditions,

    (ii) the volume of water released by storm run-off,

    (iii) localized micro-climatic conditions,

    (iv) erosion of exposed land surfaces and increase in the load of sediments and nutrients released into streams and estuaries.

By making provision for pockets of natural vegetation in urbanized or subdivided areas . . . important buffer zones are established; [we gain] stands of native vegetation as gene pools for wild species; habitats for associated fauna, including many bird species; a nutrient source in the aquatic food chain; reduction of surface run-off; reduction of erosion; the aesthetic role of providing attributes of a natural environment to the residents as part of their daily lives; areas for children to satisfy impulses to explore; visual privacy and noise abatement in residential areas; beneficial effects on the micro-climate.

Green belts:

(a) Provide shade and cool the surrounding area: an estimated 90 per cent of solar energy absorbed by trees is radiated as indirect heat and . . . significantly reduces air and ground temperatures. Air temperatures can be reduced by 3.5° and the ground temperature by much greater amounts.

    Transpiration from a single tree can absorb 633 000 kilojoules per day; which is equivalent to *five average-sized room air-conditioners running approximately twenty hours a day.*

(b) Produce oxygen. (All plants produce oxygen and absorb carbon dioxide.)

(c) Reduce pollutants: plants remove significant amounts of gaseous pollutants and particles from the air.

(d) Absorb noise: trees and shrubs are noise abaters; sound reduction is in the order of five to ten decibels for a wide belt of tall, dense trees.

(e) Provide wind breaks: strategically situated green belts would provide significant protection from wind damage.

The pattern of development must not be left to the spec. builder. Green belts should be an integral element in local land-use planning.